Advanced Inorganic Chemistry

Advanced Inorganic Chemistry
Applications in Everyday Life

Narayan S. Hosmane
Northern Illinois University

An imprint of Elsevier

Academic Press is an imprint of Elsevier
125 London Wall, London EC2Y 5AS, United Kingdom
525 B Street, Suite 1800, San Diego, CA 92101-4495, United States
50 Hampshire Street, 5th Floor, Cambridge, MA 02139, United States
The Boulevard, Langford Lane, Kidlington, Oxford OX5 1GB, United Kingdom

Library of Congress Cataloging-in-Publication Data
A catalog record for this book is available from the Library of Congress

British Library Cataloguing-in-Publication Data
A catalogue record for this book is available from the British Library

ISBN: 978-0-12-801982-5

For information on all Academic Press publications visit our
website at https://www.elsevier.com/books-and-journals

Working together
to grow libraries in
developing countries

www.elsevier.com • www.bookaid.org

Publisher: John Fedor
Acquisition Editor: Emily McCloskey
Editorial Project Manager: Sarah Watson
Production Project Manager: Anitha Sivaraj
Designer: Limbert Matthew

Typeset by TNQ Books and Journals

It is with great pleasure that I dedicate this book to my mentor and good friend, Emeritus Professor Russell N. Grimes of the University of Virginia, a pioneer in the chemistry of polyhedral boron clusters and an author and educator in the field of inorganic chemistry for more than four decades.

Narayan S. Hosmane

Contents

Biography of Author... xiii
Foreword xv
Preface.. xvii
Acknowledgments.. xix

PART 1   FOUNDATIONS: CONCEPTS IN CHEMICAL BONDING AND STEREOCHEMISTRY

CHAPTER 1 Electronic Structure: Quantum Theory Revisited 3
    1. Introduction: Why Do We Need to Know Quantum Theory?................. 3
    2. Quantum Mechanical Description of the Hydrogen Atom 3
        2.1 Quantum numbers and their significance 4
        2.2 Many electron atom... 6
        2.3 Valence—valence repulsion and term symbols 8
        2.4 Spin—orbit coupling 11

CHAPTER 2 Molecular Geometries... 15
    1. Introduction: Why Do We Need to Know Molecular Geometries or Shapes of Molecules?... 15
    2. Shapes of Molecules—Valence Shell Electron Pair Repulsion (VSEPR) Model.. 16
        2.1 The VSEPR approach.. 16
        2.2 Specific examples 19
        2.3 Other considerations 25
    3. Nonrigid Shapes of Molecules (Stereochemistry)............................... 26
        3.1 General concept 26
        3.2 Specific examples 27

CHAPTER 3 Molecular Symmetry—Part I: Point Group Assignment 31
    1. Introduction: Is It Necessary to Learn Molecular Symmetry?.............. 31
    2. Elements of Symmetry 31
        2.1 Symmetry operations 31
        2.2 Operations and elements.. 32
    3. Point Groups... 38
        3.1 Introduction... 38
        3.2 Rules for assigning point groups.. 39
        3.3 Examples... 39

CHAPTER 4 Group Theory: Matrix Representation and Character Tables...................**43**
    1. Introduction: Is It Necessary to Learn Group Theory?..........................43
    2. Other Properties of Symmetry Operations 43
        2.1 Sequential operations... 43
        2.2 Representations of a group.. 46
        2.3 Reducible and irreducible representations................................... 48
        2.4 Character tables 50
    3. Applications to Molecular Structure and Properties 51
        3.1 Application to quantum mechanics 51
        3.2 Bonding in a triangular planar structure (AX_3 with
            D_{3h} point group) 53
        3.3 Molecular orbital correlation diagram for trigonal
            planar structure (BF_3) 61
    4. Examples of Other Structures...63
        4.1 Square planar ML_4 63
        4.2 The π orbitals of the cyclopentadienide ion, $[C_5H_5]^-$ 64
    5. Molecular Spectroscopy 65
        5.1 Types of molecular motion.. 65
        5.2 Normal modes of vibration and symmetry.................................... 65
    Bibliography..69

PART 2   ADVANCED TOPICS-1: INTRODUCTION TO LIGANDS AND METAL COMPLEXES

CHAPTER 5 Ligands and *d*-Block Metal Complexes...**75**
    1. Introduction: Is It Necessary to Know About Ligands and Metal
        Complexes?...75
    2. Transition Metals.. 76
        2.1 Electronic structures and oxidation states 76
        2.2 Coordination compounds... 77
        2.3 Ligands (Lewis bases) 78
    3. Nomenclature of Coordination Compounds.. 80
    4. Isomerism in Coordination Compounds.. 84
        4.1 Coordination number $= 4$ 84

CHAPTER 6 Review of Bonding Theories for *d*-Block Metal Complexes **89**
    1. Introduction: Why Bonding Theories of Metal Complexes Are
        Important?...89
    2. Valence Bond Theory.. 90
        2.1 Coordination compounds... 90

2.2 Coordination number six 90
2.3 Coordination number four 91
3. Crystal Field Theory 92
3.1 Octahedral complexes 92
3.2 Complexes of other geometries 95
3.3 Trends in crystal field stabilization energy (Δ) 96
3.4 Predictions using crystal field theory: spin pairing
     of complexes 97
3.5 Distortions due to CFSE 98
4. Molecular Orbital Theory 99
4.1 Octahedral complexes 100
4.2 Other geometries 102
4.3 π Bonding in octahedral complexes 104
4.4 Back π bonding and the effective atomic number rule 106
4.5 Arene complexes 107
4.6 Other arene-like ligands 112
4.7 Benzene sandwich complexes 112

CHAPTER 7 Coordination Chemistry: Reaction Mechanisms and Their Influencing
Factors **115**
1. Introduction: What Makes Coordination Chemistry Interesting? 115
2. Modes of Substitution Reaction Mechanisms 116
2.1 Associative mode (a) (or an a intimate mechanism) 116
2.2 Dissociative mode (d) (or a d intimate mechanism) 117
3. Complications Involving Metal Complexes 119
3.1 Solvent competition 119
3.2 Effects of changing the other ligands on the complex 119
4. Activation Parameters 119
4.1 Enthalpy and entropy of activation 119
4.2 Activation volume, $\Delta V^{\ddagger}$ 121
4.3 Use of activation parameters in mechanistic studies 121
5. Examples of Different Coordination Numbers With
    Geometries and Factors Influencing Reaction Mechanism 123
5.1 Two- to six-coordinate complexes 123
5.2 Three-coordinate complexes 123
5.3 Four-coordinate complexes 126
5.4 Five-coordinate complexes of phosphorus and sulfur 127
5.5 Square planar complexes 129
5.6 Trans effect 132
5.7 Kinetic effect 133

5.8   Use of the trans effect in synthesis 133
5.9   Six-coordinate octahedral complexes 134
5.10   Isomerization during substitution 139
5.11   Base hydrolysis 140
5.12   Other complexes 144
5.13   Electron transfer reactions 145
Bibliography 149

PART 3   ADVANCED TOPICS-2: ELECTRONIC SPECTRA, CLUSTERS & ISOLOBAL FRAGMENTS

CHAPTER 8 Coordination Chemistry: Electronic Spectra **155**
1. Introduction: Why Do We Need to Learn Electronic Spectra? 155
2. Electronic Spectra 155
   2.1 Selection rules 155
   2.2 Spectra of octahedral (O_h) and tetrahedral (T_d) complexes 157
   2.3 Orgel diagrams 158
3. Tanabe–Sugano Diagrams 167
4. Charge Transfer Spectra 168

CHAPTER 9 Cluster Chemistry and Isolobal Fragments **171**
1. Introduction: Role of Cluster Chemistry in Nature 171
2. Clusters of Boranes, Carboranes, and Their Metal Complexes 172
   2.1 Terminology used in polyhedral boron clusters 172
   2.2 Bonding in boron clusters 174
   2.3 Nomenclature: Wade's rules and structural pattern 176
3. Clusters of Other Main Group Elements and Transition Metals 180
   3.1 Zintl anions 180
   3.2 Other main group cages 182
4. Extension of Wade's Rules Beyond Boron Clusters 182
   4.1 Mixed main group/transition metal clusters 184
   4.2 Capping groups 185
   4.3 Condensed clusters 186
   4.4 Clusters with interstitial atoms 188
   4.5 Isolobal relationships 188
   4.6 Isolobal relationships with fragments not derived from
        noble gas structures 193
Bibliography 194

PART 4   ADVANCED TOPICS-3: ORGANOMETALLIC CHEMISTRY AND CATALYSIS

CHAPTER 10 Organometallic Chemistry **199**

    1. Introduction: What Is in Organometallic Chemistry? 199

    2. Definitions and Nomenclature of Organometallic Compounds 200

        2.1 Types of organometallic compounds 200

        2.2 IUPAC nomenclature for organometallic compounds 201

    3. Molecular Formulas and Structures of Organometallic Compounds 201

        3.1 Effective atomic number rule 201

        3.2 Metal carbonyls 201

        3.3 Alkene complexes... 205

        3.4 Aromatic complexes 207

    Bibliography 208

CHAPTER 11 Catalysis With Organometallics **209**

    1. Introduction: What Is so Special About Catalytic Processes in Our Daily Life? 209

    2. Homogeneous Catalysts 209

        2.1 Oxidative addition and reductive elimination 209

        2.2 Insertion reaction (ligand migration reactions) 211

    3. Hydrogenation Catalysts... 212

        3.1 Wilkinson's catalyst, $RhCl(PPh_3)_3$ 212

        3.2 Monohydride complexes 213

        3.3 Hydroformylation and other closely related oxo processes 214

    4. Other Catalytic Processes 216

        4.1 Production of acetic acid from CH_3OH 216

        4.2 Heterogeneous catalysis of alkene polymerization with Ziegler—Natta catalyst... 217

        4.3 Suzuki—Miyaura cross-coupling reaction 218

        4.4 Sonogashira cross-coupling reaction 218

        4.5 Olefin metathesis with Grubbs and Schrock catalysts 220

    Bibliography.. 222

PART 5   ADVANCED TOPICS-4: BIOINORGANIC CHEMISTRY AND APPLICATIONS

CHAPTER 12 Bioinorganic Chemistry and Applications **225**

    1. Introduction.. 225

    2. History and Medical Relevance 225

        2.1 Salvarsan.. 226

2.2 Vitamin B12 227
2.3 Cisplatin and cancer treatments 228
2.4 Other therapeutic applications of organometallic
    compounds 231
2.5 Diagnostic metallodrugs 233
3. Transport and Storage of Metal Ions 234
3.1 Iron storage: transferrin 235
3.2 Iron storage: ferritin 236
3.3 Siderophores 237
3.4 Sodium−potassium pump 239
4. Oxygen Transport and Activation Proteins 240
4.1 Hemoglobin 241
4.2 Myoglobin 243
4.3 Hemocyanin 245
5. Biomineralization 246
Bibliography 247

Index 251

Biography of Author

Narayan S. Hosmane was born in Gokarna, Karnataka state, Southern India, and is a BS and MS graduate of Karnataka University, India. He obtained a PhD degree in Inorganic Chemistry in 1974 from the University of Edinburgh, Scotland, under the supervision of Professor Evelyn Ebsworth. After a brief postdoctoral research training in Professor Frank Glockling's laboratory at the Queen's University of Belfast, he joined the Lambeg Research Institute in Northern Ireland and then moved to the United States to study carboranes and metallacarboranes. After a brief postdoctoral work with W.E. Hill and F.A. Johnson at Auburn University and then with Russell Grimes at the University of Virginia, in 1979 he joined the faculty at the Virginia Polytechnic Institute and State University where he received a Teaching Excellence Award in 1981. In 1982 he joined the faculty at Southern Methodist University, where he became Professor of Chemistry in 1989. In 1998, he moved to Northern Illinois University and is currently a Distinguished Faculty, Distinguished Research Professor, and Inaugural Board of Trustees Professor. Dr. Hosmane is widely acknowledged to have an international reputation as "one of the world leaders in an interesting, important, and very active area of boron chemistry that is related to Cancer Research" and as "one of the most influential boron chemists practicing today." Hosmane has received numerous international awards that include but are not limited to the Alexander von Humboldt Foundation's Senior U.S. Scientist Award twice; the BUSA Award for Distinguished Achievements in Boron Science; the Pandit Jawaharlal Nehru Distinguished Chair of Chemistry at the University of Hyderabad, India; the Gauss Professorship of the Göttingen Academy of Sciences in Germany; Visiting Professor of the Chinese Academy of Sciences for International Senior Scientists; High-End Foreign Expert of SAFEA of China; and Foreign Member of the Russian Academy of Natural Sciences. He has published over 325 papers in leading scientific journals and is an author/editor of five books on Boron Science, Cancer Therapies, General Chemistry, Boron Chemistry in Organometallics, Catalysis, Materials and Medicine, and this book on Advanced Inorganic Chemistry.

Foreword

It is a truism that chemistry is a moving, ever-changing stream, a fact well known not only to chemists but also to anyone with even a passing interest in the subject. One need only compare the journal publications of today with those of just a few years ago in the same field to realize the astonishing rapidity of movement on the scientific frontiers. In my lifetime I have seen entire fields of study arise (often from a single discovery), grow, thrive, decline, revive, or seem to disappear, only to rise again propelled by an unanticipated finding. Yet the *teaching* of chemistry evolves much more slowly, as reflected in course content and textbooks. College-level treatments of basic chemistry typically change only incrementally from year to year, with new discoveries dutifully noted but with little alteration in the layout of the courses; class notes used by instructors may endure for years or decades. Advanced courses for upper level undergraduates and graduate students are more likely to reflect new developments, but at this level the enthusiasm of students is usually so high that even moderately gifted professors can enjoy success. The real challenge, as I found in decades of university teaching, is found in the general, organic, and physical chemistry courses required for a BS or BA degree, which are populated by captive audiences who see the material as an endurance test and the professor as a drill sergeant. It is to this group that Professor Hosmane directs this book. In this innovative text, he presents an approach that seeks to engage students' interest by asking, in effect, "Why do I need to know this? What good is it?" The mere suggestion that there is a real purpose—a method to the madness, as it were—beyond the dissemination of knowledge for its own sake, is likely to raise eyebrows and stimulate real interest in the material. Each new topic is introduced by explaining its relevance, indeed its fundamental importance, to biochemistry and other relevant areas, in a way that is more likely to capture the reader's attention than does a more pedantic and traditional style. Students are especially likely to embrace this approach, and this text is a welcome new tool for teaching the centuries-old, yet constantly evolving, field of inorganic chemistry.

Russell N. Grimes
Emeritus Professor of Chemistry
University of Virginia

Preface

The lack of connectivity between the topics we read about and what we experience in nature has been a fundamental drawback in any textbook. No wonder, inorganic chemistry has been a nightmare subject for many students and the instructors. Therefore, I had to teach the subject from a totally different angle! For example, I wanted my students to learn the shapes (geometry) dictating the intermolecular forces of attractions which influence the reaction between molecules of different shapes. In turn, the reactivity leads to complex formation via a number of mechanisms (associative, dissociative, interchange associative, and interchange dissociative, etc., with the continuous classroom exit and entrance versus entrance into an empty classroom as examples) and how the coordination chemistry between the transition metals and the ligands has a direct correlation with cyanide or carbon monoxide poisoning [strong-field cyanide (CN) or carbon monoxide (CO) ligand versus weak-field oxygen (O_2) molecule] that could make sense to the biochemistry majors who are not aware of the connectivity between inorganic chemistry and biochemistry despite the subject being required for ACS accreditation for the BS degree graduation! Similarly, the applications of organometallic chemistry, catalysis, cluster chemistry, and bioinorganic chemistry in producing durable polymeric materials, drugs, etc., are directly correlated with what we see and experience in our daily lives. Therefore, I have written this new textbook on advanced inorganic chemistry with simple explanations of these concepts relate them to things we see and experience in nature. Perhaps this approach might rekindle, in an agreeable way, the interest of the students in learning this subject, which they may have thought to be uninteresting.

Narayan S. Hosmane

Acknowledgments

In the preparation of this manuscript, several individuals have been unusually helpful, especially Ms. Lauren Zuidema, Mr. Lucas Kuzmanic, and Dr. P. M. Gurubasavaraj (Visiting Raman Fellow from India). Chapter 12 is taken almost directly from Lauren and Lucas's research paper entitled "Bioinorganic Chemistry and Applications."

Ms. Lauren Zuidema   Mr. Lucas Kuzmanic   Dr. P. M. Gurubasavaraj

It has been modified to fit into the format of this book, although I tried to maintain as much possible the effectiveness of Lauren and Lucas's original writing. Dr. P. M. Gurubasavaraj oversaw the work of these two young researchers; I express my sincere thanks to Dr. Gurubasavaraj for this help. I am grateful to Mr. Hiren Patel, an artist of exceptional caliber, for his help in creating the cover page for the book. My special thanks go to Dr. Yinghuai Zhu and Professor Dennis N. Kevill of Northern Illinois University who kindly agreed to read the manuscript and made invaluable suggestions.

Last, but not least, I wish to express my thanks to Acquisitions Editor Katey Birtcher, and Senior Editorial Project Manager Jill Cetel of Elsevier Publishing Inc. for their continuous support and patience. If it were not for Katey's persuasive ability, I would not have committed to this venture, and, in turn, would not have attempted to persuade my longtime collaborator Professor John Maguire of Southern Methodist University into joining me in this venture, even though unsuccessfully.

Narayan S. Hosmane

Part **1**

Foundations: Concepts in Chemical Bonding and Stereochemistry

Special Emphasis on the Symmetry Groups, Molecular Symmetry, and Molecular Orbital Theory

ABSTRACT

This section of Inorganic Chemistry builds upon the general knowledge gained in the freshman chemistry course. Specifically, the chapters introduce the geometrical shapes that are classified into symmetry groups by describing their molecular symmetry and their important roles in chemical reactivities. All of this information leads to the ultimate theoretical interpretation called "Molecular Orbital Theory," without which it would have been impossible to predict and then explain why some molecules are "paramagnetic," which allows these molecules to be useful in magnetic resonance imaging for cancer diagnosis. Furthermore, it was this theory that led to our modern day information technology, involving materials that are semi- and superconductors. Therefore, it is imperative to strengthen our foundation of knowledge before exploring other advanced areas of Inorganic Chemistry.

1

Foundations: Concepts in Chemical Bonding and Stereochemistry

ABSTRACT

Electronic Structure: Quantum Theory Revisited

1. INTRODUCTION: WHY DO WE NEED TO KNOW QUANTUM THEORY?

Whenever the word "Quantum" is introduced, the first thing that comes to anyone's mind is "Physics and the Laws". These govern the human approach to a study of the universe as introduced in the 20th century. "Quantum" is the word used for the smallest scale of any discrete object. A tiny "bundle" involved in radiant energy is equal to the multiplication of Planck's constant (h) with the frequency (v) of the associated radiation. Thus, Max Planck's discovery of "black-body radiation" in 1900 combined with Albert Einstein's experiment in 1910 of "photoelectric effect" gave the first explanation and application of "Quantum Theory". These led to the discovery of "line spectra" to describe Niels Bohr's model of the atom with quantized orbits in 1913, followed by Louis de Broglie's discussion of "wave-particle duality" in 1923, combined with Heisenberg's Uncertainty Principle in 1927 and Erwin Schrödinger's approximation in 1926 to locate the position of electrons in an atom through his partial differential equation for the wave functions of particles. While the uncomplicated Newton's laws, when applied to thermal physics, failed to explain the unusual properties of the subatomic particles, the modern atomic theory through quantum mechanics succeeded beyond imagination, and this is exactly the reason why we should study the quantum theory, so that we can consider the mysteries of nature.

2. QUANTUM MECHANICAL DESCRIPTION OF THE HYDROGEN ATOM

Using the orbitals of the hydrogen atom with its associated energies, one can construct approximations for any molecule with more complex wave functions.

Advanced Inorganic Chemistry. http://dx.doi.org/10.1016/B978-0-12-801982-5.00001-1

2.1 **Quantum numbers and their significance**

1. Wave function (the Greek letter "psi") for H-like atom in Schrödinger's equation:

$$H\Psi = E\Psi$$

$$\Psi_{(n,l,m_l)} = \text{Orbital} = Y_{(l,m_l)}R_{(n,l)}$$

Y = angular part of wave function, R = radial part of wave function, $|\Psi|^2 \propto$ probability density

2. Significance of quantum numbers

   a. n = principal quantum number

   $n = 1, 2, 3, 4, 5 \ldots$ any integer number, important in specifying the energy of electron and the radial distribution function, $P_r = 4\pi r^2 R^2_{(n,l)}$. The most probable and average value of r increases as n increases.

   b. l = Azimuthal quantum number

   i. $l = 0, 1, 2, 3, 4 \ldots (n-1)$

   ii. Orbital angular momentum M

   Ψ is an eigen function of M^2 in that $M^2\Psi = l(l+1)\hbar^2\Psi$
   Total orbital angular momentum $= \sqrt{l(l+1)}\hbar$ ($\hbar$ is the reduced Planck constant)

   iii. Energy of electrons depends on both n and l
   Recall subshell notations 1s ($n = 1, l = 0$), 3d ($n = 3, l = 2$)

   c. Magnetic quantum number, m_l

   i. Depends on value of l. $m_l = l, l-1, l-2, \ldots 0 \ldots -l$. total number $= 0, \pm 1, \pm 2, \pm l$

   ii. There are $2l + 1$ values possible for m_l. In the absence of magnetic fields, each l state is $2l + 1$ fold degenerate.

   iii. m_l specifies the "z" component of the electron's orbital angular momentum.
   Ψ is an eigen function of the "z" of M_z = operator for "z" component of orbital angular momentum

   $$M_z\Psi = m_l\hbar\Psi$$

   iv. Ψ is not an eigen function of either M_x or M_y. The average values of the "x" and "y" components of the orbital angular momentum $= 0$.

3. Vector model of atom angular momentum
   a. Orbital angular momentum acts as a vector of magnitude $\sqrt{l(l+1)}\hbar$ that precesses about the "z" axis and has a projection along the "z" axis of $m_z\hbar$ (Fig. 1.1).

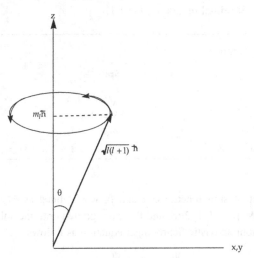

■ **FIGURE 1.1** Vector model of atom angular momentum.

   b. Magnetic Properties. Since an electron is charged, its orbital motion will generate a magnetic moment, μ.

$$\mu = \frac{\mu^0}{\hbar} \cdot (\text{angular momentum})$$

$\therefore \mu = $ total orbital magnetic moment $= \sqrt{l(l+1}\,\mu^0$

$\mu_z = $ "z" component of magnetic moment $= m_l\mu^0$

$\mu^0 = $ Bohr magnetron $= 9.27 \times 10^{-24} \text{ JT}^{-1}$

4. Electron spin—intrinsic properties of electrons
   a. From atomic spectroscopy and magnetic measurements, it became apparent that individual electrons possess an intrinsic angular momentum of $1/2\hbar$ and an intrinsic magnetic moment of μ° oriented either parallel or antiparallel to its orbital momentum and magnetic moment. The origin of these properties is relativistic, but we will use the term "electron spin" when referring to them. Since these properties do not significantly affect the energy of the electron and

those effects arise from these properties, they can be accounted for by adding specific terms to the kinetic energy and potential energy parts of the Hamiltonian. We will use an approach proposed by Pauli.

b. We will define a set of spin functions and operators to parallel those of orbital motion (Table 1.1).

Table 1.1 Spin Functions and Orbital Motion Operators

	Orbital	Spin
Total angular momentum	$\sqrt{l(l+1)}\hbar$	$\sqrt{s(s+1)}\hbar$ $S = 1/2$
"z" component	$m_l\hbar$ $(m_l = l, l-1... -l)$	$m_s\hbar$ $(m_s = s, s-1... -s) = 1/2, -1/2$

c. A set of spin functions, α and β, was defined as $M_s\alpha = 1/2\hbar\alpha$ and $M_s\beta = -1/2\hbar\beta$, and these are grafted onto the solutions for the nonrelativistic Schrödinger equation as follows.

$$\Psi_{(n,l,m_l)}\alpha = \Psi_{\left(n,l,m_l,\frac{1}{2}\right)}$$

$$\Psi_{(n,l,m_l)}\beta = \Psi_{\left(n,l,m_l,-\frac{1}{2}\right)}$$

2.2 **Many electron atom**

Because of the electron–electron term in the potential energy term, the Schrödinger equation cannot be exactly solved for a many (more than one) electron atom and one must approximate. Zero-order approximation is just to ignore the terms that are smaller than the electron-nuclear attraction terms. In that case, for an atom with N electrons:

1. $\Psi = \prod_{i=1}^{N} \psi_i$ and $E = NE_H$. These are not good solutions but are useful starting points.

2. Pauli's exclusion principle—there are two ways to state:

   a. No two electrons in the same atom can have all four quantum numbers the same, and two electrons in the same orbital must have their spins paired.

   b. The total wave function must be antisymmetric to the interchange of electrons. If P is an operator that interchanges two electrons (permutation operator) then $P\Psi^2 = \Psi^2$ or $P\Psi = \pm\Psi$. + sign means symmetric and − sign means unsymmetric. Therefore, Ψ must change sign on permutation of electrons.

c. Consider the case where N = 2 (He atom). Possible wave functions $\Psi = \phi_{1s}\alpha(1)\phi_{1s}\beta(2)$; however, $P\Psi = \phi_{1s}\alpha(2)\phi_{1s}\beta(1)$ and it is neither symmetric nor antisymmetric.

$$\Psi = \frac{1}{\sqrt{2}}[\phi_{1s}\alpha(1)\phi_{1s}\beta(2) + \phi_{1s}\alpha(2)\phi_{1s}\beta(1)] \text{ symmetric}$$

$$\Psi = \frac{1}{\sqrt{2}}[\phi_{1s}\alpha(1)\phi_{1s}\beta(2) - \phi_{1s}\alpha(2)\phi_{1s}\beta(1)] \text{ antisymmetric}$$

We can ensure an antisymmetric wave function by using a determinantal form:

$$\Psi_{He} = \frac{1}{\sqrt{2}}\begin{vmatrix} \phi_{1s}\alpha(1) & \phi_{1s}\beta(1) \\ \phi_{1s}\alpha(2) & \phi_{1s}\beta(2) \end{vmatrix}$$

2.2.1 Effects of electron–electron repulsion

1. The effects can be cataloged in two ways.

   Electrostatic—The electrons will shield the nuclear charge as seen by other electrons.

   Electron correlation—the motion of one electron will affect the motion of the other electrons.

2. Classification of electrons

   Core electrons. Electrons in shell which have a lower main quantum number n. They are in spherically symmetric closed shells and are chemically inert (most of the time). Their main function is to shield the nuclear charge as seen by the valence electrons.

   Valence electrons. Outermost electrons which are frequently in partially filled subshells. They are chemically and spectroscopically active.

3. Electrostatic effects—we can use the Hartree-Foch method.

   Assume that each electron moves in an average field due to the nucleus and other electrons. If Ψ's are known, then those can be used to calculate the average field and solve the Schrödinger equation. In the absence of Ψ's, one cannot solve the Schrödinger equation due to cyclic problem. Therefore, assume a set of Ψ's (hydrogen-like anti-symmetric functions); use these Ψ's to solve the Schrödinger equation to get a new, better set of Ψ's. Repeat with the new Ψ's to get an even better set. Continue this process until the Ψ's you get are essentially indistinguishable from those you put in. In that case, you have reached a self-consistent field.

The following results can be expected.

The sequence of energy levels encountered follows the aufbau principle:

1s < 2s < 2p < 3s < 3p < 4s < 3d < 4p < 5s < 4d < 5p < 6s < 4f ≈ 5d < 6p < 7s ...

One can write electron structures by writing electron configurations where (1) energy levels change as atomic number increases and thus the above sequence gives the next level that is encountered when auf-bau principle is followed, (2) one cannot distinguish a single electron. Due to electron exchange, only total electron density must be considered, and (3) $Y(l,m_l)$ is not changed from H and $R(n,l)$ is different.

2.3 Valence—valence repulsion and term symbols

1. General

Consider carbon. The electron configuration is $1s^2 2s^2 2p^2$.

a. Number of wave functions possible for the p^2 configuration.

There are six one-electron orbitals, each distinguishable by different m_l and m_s.

$$+1\alpha \quad +1\beta \quad 0\alpha \quad 0\beta \quad -1\alpha \quad -1\beta$$

In the absence of strong magnetic fields, they all have the same energy.

$$\text{No. of combinations} = \frac{6 \times 5}{2} = 15$$

b. What are the characteristics of the different states and which is the lowest energy state?

First, we will use term symbols to identify the states, and then we will use Hund's rule to select the lowest energy state (ground state).

2. Term Symbols

a. Designate each state by its orbital angular momentum and spin angular momentum. Use the vector model to determine these quantities.

b. Orbital angular momentum—Recall that each electron has an orbital angular vector of $\sqrt{l_i(l_i+1)}\hbar$, which is no longer constant with time due to the presence of the other electrons (electron—electron correlation).

The individual vectors will add (vectorially) to give a resultant **total orbital angular momentum** vector of $\sqrt{L(L+1)}\hbar$ that is constant and precesses about the "z" axis.

  i. L = total orbital angular momentum quantum number.
  ii. Electrons in filled subshells contribute zero to L.

iii. Can deduce L by determining the different values of the "z" component of the total orbital angular momentum, $M_L\hbar$

$M_L = L, L-1, L-2, L-3 \ldots -L \therefore$ each L state is $2L+1$ fold degenerate.

M_L can be obtained directly from the individual m_{li}

$$M_L = \sum m_{l_i}$$

Once M_Ls are known, the Ls can be determined.

c. Spin Angular Momentum. Because of electron correlation, the individual spin angular momentum vectors, $\sqrt{s_i(s_i+1)}\hbar$ ($s_i = 1/2$), are no longer constant with time.

The individual spin vectors will add to give a **total spin angular momentum** vector of $\sqrt{S(S+1)}\hbar$ that precesses about the "z" axis.

i. S = total spin angular momentum quantum number

ii. The "z" component of the spin angular momentum = $M_s\hbar$

$$M_S = S, S-1, S-2\ldots -S \quad M_s = \sum m_{s_i}$$

Each S state is $2S+1$-fold degenerate = **multiplicity** (Table 1.2).

Table 1.2 Spin Multiplicity and Labeling		
S	**Multiplicity (2S + 1)**	**Label**
0	1	Singlet
1/2	2	Doublet
1	3	Triplet
3/2	4	Quartet

iii. The values of S can be determined from the number of possible unpaired electrons in accordance with the **Pauli's exclusion principle**.

If N = maximum number of unpaired electrons, then

$$S = \frac{N}{2}, \frac{N}{2}-1, \frac{N}{2}-2\ldots \begin{cases} 0 \text{ (if N is even)} \\ \frac{1}{2} \text{ (if N is odd)} \end{cases}$$

iv. Example: If N = 3, $S = \frac{3}{2}, \frac{1}{2}$

If $S = \frac{3}{2}$, $2S+1 = 4$, Quartet State

Combinations : $\quad \alpha\alpha\alpha \quad \alpha\alpha\beta \quad \alpha\beta\beta \quad \beta\beta\beta$

$\qquad\qquad M_{\overline{S}} = \quad 3/2 \quad 1/2 \quad -1/2 \quad -3/2$

If $S = \frac{1}{2}, 2S + 1 = 2$, Doublet State

$$\text{Combinations}: \quad \alpha\beta\alpha \quad \beta\alpha\beta \quad \text{or} \quad \beta\alpha\alpha \quad \beta\beta\alpha$$
$$M_s = \quad 1/2 \quad -1/2 \quad\quad 1/2 \quad -1/2$$

d. Term Symbols

    i. Term symbol gives the values of L and S for the energy state of a many electron atom. Use letter designation to specify L

$$L = 0 \quad 1 \quad 2 \quad 3 \quad 4 \quad\quad\quad \cdots$$
$$\quad\quad S \quad P \quad D \quad F \quad G \quad \text{further using letters of the alphabet}$$

Multiplicity can be: $2S + 1$, as a left-hand superscript. L^{2S+1} Total degeneracy of the state $= (2L + 1)(2S + 1)$. Example can be given as: ^3P (triplet P state) $L = 1$, $S = 1$

    ii. Table 1.3 shows the states arising from the p^2 configuration. The M_L and M_S values for all 15 micro-states can be written in a long way, and L and S can be deduced from them.

Have a ^1D state where $L = 2$; $S = 0$. Degeneracy $= (4+1)(0+1) = 5$

Have a ^3P state where $L = 1$; $S = 1$. Degeneracy $= (2+1)(2+1) = 9$

Have a ^1S state where $L = 0$; $S = 0$. Degeneracy $= \underline{(0+1)(0+1) = 1}$

                                  Total Degeneracy $= 15$

Short way: Determine values of S from the number of unpaired electrons possible, then get possible values of M_L for each using the Pauli's exclusion principle.

$N = 2$ ∴ $S = 1$, 0 have triplet (Table 1.4) and singlet states (Table 1.5).

Table 1.3 All States Arising From p^2 Configuration

m_l	m_s	1	2	3	4	5	6	7	8	9	10	11	12	13	14	15
1	1/2	X	X	X	X	X										
1	−1/2	X					X	X	X	X						
0	1/2		X				X				X	X	X			
0	−1/2			X				X			X			X	X	
−1	1/2				X				X			X		X		X
−1	−1/2					X				X			X		X	X
M_L		2	1	1	0	0	1	1	0	0	0	−1	−1	−1	−1	−2
M_S		0	1	0	1	0	0	−1	0	−1	0	1	0	0	−1	0
State		^1D	^3P	^1D	^3P	^1D	^3P	^3P	^3P	^3P	^1S	^3P	^1D	^3P	^3P	^1D

Table 1.4 Triplet State, $m_{l_1} \neq m_{l_2}$

$m_l(1)$	$m_l(2)$	M_L
1	0	1
1	−1	0
0	−1	−1

$L = 1. \therefore {}^3P$ *state.*

Table 1.5 Singlet State, $m_{l_1} = m_{l_2}$

$m_l(1)$	$m_l(2)$	M_L	State
1	1	2	${}^1D$
1	0	1	${}^1D$
1	−1	0	${}^1D$
0	0	0	${}^1S$
0	−1	−1	${}^1D$
−1	−1	−2	${}^1D$

$L = 2, 0. \therefore {}^1D$ *and* ${}^1S$.

  e. Hund's rules for the ground state
   i. The state with the highest multiplicity lies lowest.
   ii. Of those states with the same multiplicity, the one with the largest L is the lowest in energy.
   iii. Ground state is ${}^3P$. Hund's rules only allows for the selection of the ground state and cannot be used to order the energy states. The complete sequence must be determined from atomic spectroscopy. For p^2, the order is: ${}^3P < {}^1D < {}^1S$

2.4 Spin—orbit coupling

1. Spin—orbit coupling is an effect in addition to the electron—electron repulsion effect.
  a. It occurs due to the interaction of the magnetic moment generated and the intrinsic moment of the electron.
  b. Must add a new term to the Hamiltonian operator, $H_{so} = \sum_{i=1}^{N} \varepsilon_i \overrightarrow{M_{l_i}} \cdot \overrightarrow{M_{s_i}}$ $\varepsilon_i = $ spin—orbit coupling constant, which increases as the atomic number increases.
   i. For low atomic numbers, one can use an approximation called the Russell-Saunders or $L \bullet S$ coupling.

    ii. Assume L and S are still good quantum numbers. The total orbital angular momentum vector, $\sqrt{L(L+1)}\hbar$, and the total spin angular momentum vector, $\sqrt{S(S+1)}\hbar$, will add vectorially to give a **Total Angular Momentum** vector of $\sqrt{J(J+1)}\hbar$ that precesses about the "z" axis.

    iii. The "z" component $= M_J\hbar$ $(M_J = J, J-1, J-2 \ldots -J)$. Each J state is $2J+1$ fold degenerate.

    iv. $J = L+S, L+S-1, L+S-2 \ldots |L-S|$ J is written as a right-hand subscript to the term symbol.

    Example: ^3P state $(L=1, S=1)$; therefore $J = 2, 1, 0$

$$^3P \begin{cases} ^3P_0 \text{ degeneracy} = 1 \ (2\,J{+}1) \\ ^3P_1 \text{ degeneracy} = 3 \\ ^3P_2 \text{ degeneracy} = 5 \end{cases}$$

  c. Hund's fine structure rule.

    i. In atoms with less than half-filled subshells, the lowest value of J lies lowest overall.

    ii. In atoms with more than half-filled subshells, the highest value of J lies lowest. Therefore, for a p^2 configuration, the $^3P_0$ state is the ground state.

2. In atoms with a large atomic number, the spin orbit term becomes large compared to the electron–electron repulsion term and the simple L-S coupling scheme does not work. In this case, a type of coupling called j–j coupling is operative.

3. Effects of Spin–orbit coupling

  a. Fine structure in atomic spectra. The principal Na spectral line is the "D doublet" at wave lengths of 589.76 and 589.16 nm. Due to the transition from a $3p^1$ to a $3s^1$ state, p^1 gives rise to ^2P and s^1 to a ^2S. However, due to spin–orbit coupling, the ^2P is split into two states, a $^2P_{3/2}$ and a $^2P_{1/2}$. The ^2S is just $^2S_{1/2}$. The resulting transitions can be seen in Fig. 1.2

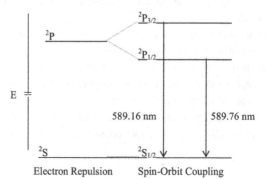

■ **FIGURE 1.2** Correlation diagram between electron repulsion and Spin–orbit coupling.

b. Anomalous Zeeman Effect

    i.   No unpaired electrons ($S = 0$, $J = L$). In a magnetic field, the degeneracy with respect to M_L is increased and one gets more spectral lines (Zeeman effect).

    ii. Unpaired electrons ($S \neq 0$, $J \neq L$). In magnetic field, M_J degeneracy is increased and one gets increasingly more complex spectra (anomalous Zeeman effect).

Chapter **2**

Molecular Geometries

1. INTRODUCTION: WHY DO WE NEED TO KNOW MOLECULAR GEOMETRIES OR SHAPES OF MOLECULES?

Whenever a question is posed to a student of biochemistry about "why do we need to know about shapes or geometry of molecules?", the immediate answer is that it is a waste of time for them to know, as it does not matter much in their future career as a professional in medical sciences or in pharmaceutical industries. When it is emphasized that the shapes or geometries play a vital role in natural life processes that occur daily in every living thing throughout nature, the students begin to wonder how this could be part of nature. Just take any shape or geometry that we know or read about and try to fit it in a circle, and then imagine the universe as a big circle. After this simple experiment, we all agree that nature created all shapes and sizes, just like each molecule or matter comes in various sizes and shapes. We all learn that intermolecular forces of attraction are due to polarity between the molecules, and this polarity is dictated by their shapes. In other words, shape leads to attraction between the two polar ends, just like the intermolecular forces of attractions, such as ion—ion, ion—dipole, dipole—dipole, and even van der Waal forces among neutral species leading to induced dipole—dipole. It is clear now that the shapes or geometries with their proper orientations in 3D dictate intermolecular forces of attraction leading to reactivity between the molecules that will yield products. These facts of nature are exactly the reason why we need to learn all geometrical shapes and symmetries, and how important are these in predicting the reactivity patterns, whether it may be within our bodies, outside in the backyard, or in the universe. Since Part 1 is all about the foundation of knowledge dealing with classification of symmetry groups, molecular symmetry, and molecular orbital theory, we will first refresh our general chemistry knowledge on all kinds of geometrical shapes. Thus, Chapter 2 will discuss molecular geometries.

Advanced Inorganic Chemistry. http://dx.doi.org/10.1016/B978-0-12-801982-5.00002-3

2. SHAPES OF MOLECULES—VALENCE SHELL ELECTRON PAIR REPULSION (VSEPR) MODEL

2.1 The VSEPR approach

The primary approach to predict the shape of any molecule is to follow the VSEPR model with the following sequence of steps.

1. Draw a Lewis diagram.
2. Count the number of **lone pairs (L)** + **bonded atoms (B)** around each central atom. At this point, it does not matter whether the atoms are bonded by single, double, or triple bonds.
3. These groups of bonded atoms and lone pairs will repel one another and arrange themselves about the central atom so that they are as far away from one another as possible. The order of repulsion is: lone pairs >> triple bonds > double bonds > single bonds.
4. If the sum of L + B about a central atom is equal to
   a. 2, the arrangement is **Linear** (Fig. 2.1).

■ **FIGURE 2.1** L + B = 2, linear geometry.

   b. 3, the arrangement is **Trigonal Planar** (Fig. 2.2).

All three sites are equivalent

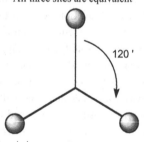

120 '

■ **FIGURE 2.2** L + B = 3, trigonal planar geometry.

c. 4, the arrangement is **Tetrahedral** (Fig. 2.3).

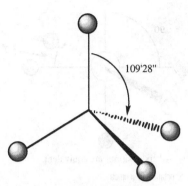

All four positions are equivalent.

■ **FIGURE 2.3** L + B = 4, tetrahedral geometry.

d. 5, the arrangement is **Trigonal Bipyramidal** (Fig. 2.4).

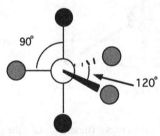

1) All five positions are not equivalent.
2) The three positions in the trigonal plane
   are the **equatorial** positions.
3) The two above and below the equatorial
   plane are the **axial** positions.

■ **FIGURE 2.4** L + B = 5, trigonal bipyramidal geometry.

e. 6, the arrangement is **Octahedral** (Fig. 2.5).

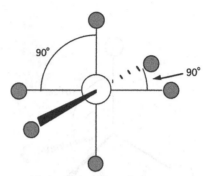

All six positions are equivalent.

■ **FIGURE 2.5** L + B = 6, octahedral geometry.

A less common geometry for six-coordinate substances is that of a **Trigonal Prism** (Fig. 2.6).

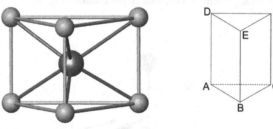

■ **FIGURE 2.6** Geometry of a trigonal prism.

Certain bidentate molecules (ligands) of the type $[S_2C_2R_2]^{2-}$ or $[Se_2C_2R_2]^{2-}$, which have a small bite angle (shorter binding distance due to single atom linker unit), can force this geometry. An example is Mo $[Se_2C_2(CF_3)_2]_3$. This geometry is thought to be present in an important intermediate structure in the intramolecular rearrangement of some octahedral complexes.

2.2 **Specific examples**

1. B + L = 2; Linear (Fig. 2.7).

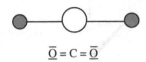

$$\overline{O} = C = \overline{O}$$

Example: H–C≡C–H

■ **FIGURE 2.7** Example of linear geometry, L + B = 2.

2. L + B = 3

Lone pairs in this theory behave in the same way as bonded atoms in determining geometries (Fig. 2.8).

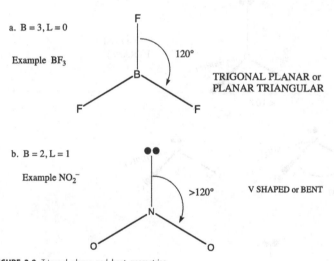

a. B = 3, L = 0

Example BF₃

120°

TRIGONAL PLANAR or
PLANAR TRIANGULAR

b. B = 2, L = 1

Example NO₂⁻

>120°   V SHAPED or BENT

■ **FIGURE 2.8** Trigonal planar and bent geometries.

3. $L + B = 4$; Tetrahedral (Fig. 2.9).

a. $B = 4, L = 0$ CH_4, SO_4^{2-}, NH_4^+

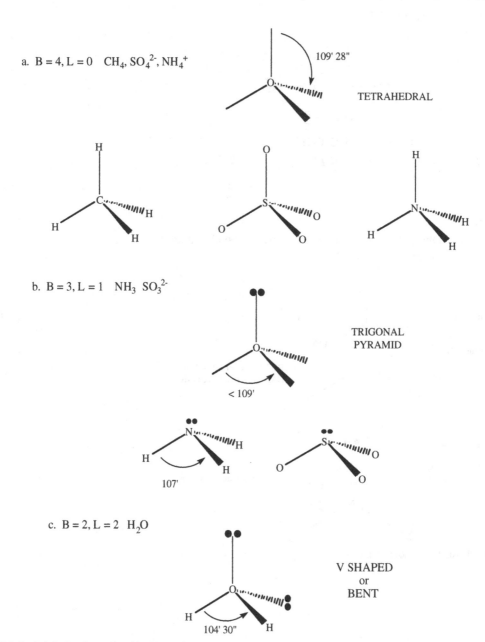

TETRAHEDRAL

b. $B = 3, L = 1$ NH_3 SO_3^{2-}

TRIGONAL
PYRAMID

c. $B = 2, L = 2$ H_2O

V SHAPED
or
BENT

■ **FIGURE 2.9** Tetrahedral, trigonal pyramid, and bent geometries.

4. L + B = 5; Trigonal Bipyramid (Fig. 2.10).

a. $B = 5, L = 0$ PF_5

The F's are not equivalent. The two axial bonds are 219 pm and the three equatorial bonds are 204 pm

b. $B = 4, L = 1$ SF_4

c. $B = 3, L = 2$ IF_3 "T" SHAPED

d. $B = 2, L = 3$ I_3^-

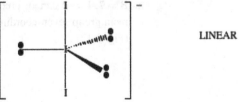   LINEAR

■ **FIGURE 2.10** Trigonal bipyramid, seesaw, T-shaped, and linear geometries.

5. $L + B = 6$; Octahedral (Fig. 2.11).

a. B = 6, L = 0   SF$_4$

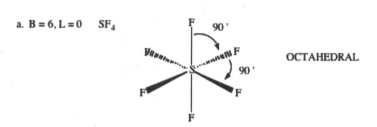

OCTAHEDRAL

b. B = 5, = 1   IF$_5$

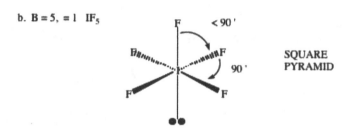

SQUARE
PYRAMID

c. B = 4, L = 2   XeF$_4$

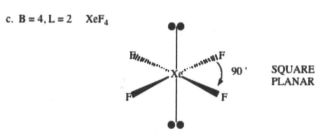

SQUARE
PLANAR

■ **FIGURE 2.11** Octahedral, square pyramid, and square planar geometries.

6. $L + B = 7$

   a. $B = 7$, $L = 0$ cannot predict a single structure. There are very few main group seven-coordinate complexes. One of the few is IF$_7$ (Fig. 2.12).

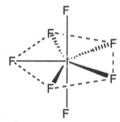

■ **FIGURE 2.12** Pentagonal bipyramidal geometry of IF$_7$.

The Nb in $K_2[NbF_7]$ is seven-coordinate but the ion is a capped trigonal prism (Fig. 2.13).

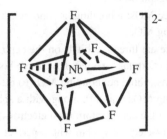

■ **FIGURE 2.13** Capped trigonal prism geometry of $[NbF_7]^{2-}$.

b. B = 6, L = 1

No simple theory can accurately predict the structure, since different compounds have different structures. The VSEPR theory predicts a distorted octahedral structure for XeF_6 (Fig. 2.14).

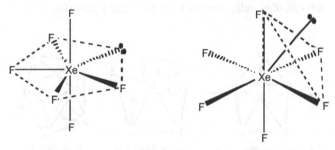

Distorted Octahedral   Capped Octahedral

■ **FIGURE 2.14** Two possibilities of distorted octahedral geometry.

Theoretical study indicates that the first structure has a slightly lower energy than the second one. Similarly, SeX_6^{2-} and TeX_6^{2-} each have B = 6, L = 1 (14 electron count), but they have perfect octahedral symmetry. We can use modified molecular orbitals to explain the structure of these molecules (Fig. 2.15).

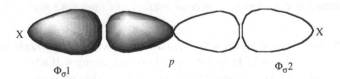

$$\Psi_{MO} = c_1\phi_p + c_2\left(\phi_{\sigma1} + \phi_{\sigma2}\right)$$

■ **FIGURE 2.15** Linear combination of *p*-orbitals on the central atom.

    i. Take a linear combination of a *p* orbital on the central atom and two ligand orbitals.

    ii. Select three MOs: a bonding, a nonbonding, and an antibonding MO.

    iii. Since there are three *p* orbitals on the central atom and six ligand σ orbitals, we will have three bonding MOs (6 electrons), three nonbonding MOs (6 electrons), and three antibonding MOs (vacant), along with a spherically symmetric *s* orbital on the central atom (two electrons). Therefore, there will be a 14-electron system with a perfect octahedral geometry.

7. Higher coordination numbers exist but are generally rare, except in the lanthanides where coordination numbers of 8, 9, 10, and 12 are known.

    a. The most common structures for coordination number 8 are the bicapped trigonal prism (BTP) $[ZrF_8]^{4-}$, square antiprism (SAP) $[Zr(acac)_4]$ (acac = acetylacetonate $[CH_3(=O)C-(CH)-C(=O)CH_3]^-$), and the dodecahedron (DD) $[Mo(CN)]^{4-}$. All can be thought of as being derived from distorting a cube (Fig. 2.16).

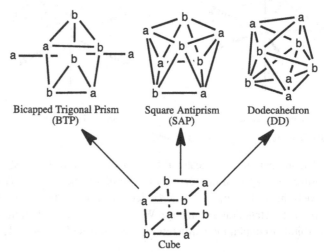

    Bicapped Trigonal Prism   Square Antiprism   Dodecahedron
    (BTP)   (SAP)   (DD)

    Cube

■ **FIGURE 2.16** Most common geometries for coordination number 8 derived from a distorted cube.

    b. Coordination numbers 9, 10, and 12 are found in compounds such as $[Ln(H_2O)_9][(BrO_3)_3]$ (tricapped trigonal prism), $K_2[Er(NO_3)_5]$ (two O atoms of each $[NO_3]^-$ coordinate in a bicapped square antiprism) and the anion in $[Mg(H_2O)_6]_3[Ce(NO_3)_6]_2.6H_2O$ (two O atoms of each $[NO_3]^-$ coordinate in an icosahedron) as shown in Fig. 2.17.

Tricapped Trigonal   Bicapped Square   Icosahedron
Prism   Antiprism

■ **FIGURE 2.17** Geometries derived from coordination numbers 9, 10, and 12.

2.3 **Other considerations**

1. Electronegativity differences.
   a. Electron—electron repulsion increases as the electrons are closer to the central atom. The electron density in a bond formed between the central atom and a very electronegative atom will be polarized away from the central atom and will be less effective in repelling electrons than when the central atom is bonded to an element with lower electronegativity.
   b. Therefore, bond angles involving very electronegative groups will tend to be smaller than those involving less electronegative groups. Also note that the presence of the π bond will cause an additional repulsion, because of its four electrons resulting in a larger bond angle, and with a smaller bond angle involving the two attached atoms.
   c. Examples are COF_2 and CH_2O (Fig. 2.18).

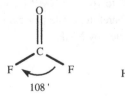

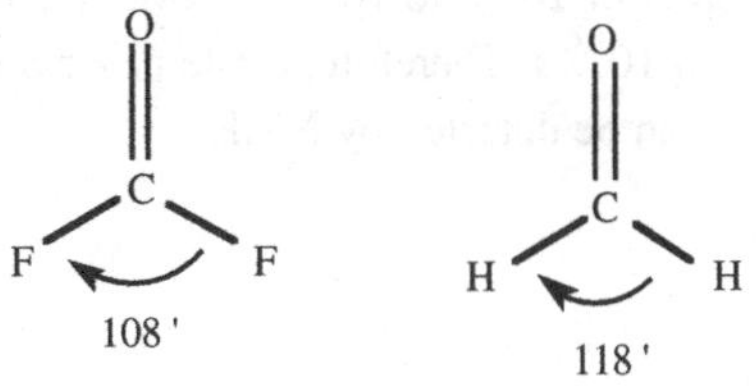

■ **FIGURE 2.18** Structures of COF_2 and CH_2O.

2. When lone pairs are present, the bond angles involving atoms decrease as the size of the central atom increases as shown in Fig. 2.19.

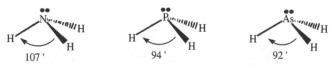

■ **FIGURE 2.19** Geometries of NH_3, PH_3, and AsH_3.

Note that the perfect octahedral symmetries of SeX_6^{2-} and TeX_6^{2-} are further examples of the nonstereochemically active s orbital in the heavier main group complexes (the so-called inert pair effect).

3. NONRIGID SHAPES OF MOLECULES (STEREOCHEMISTRY)

3.1 General concept

We are used to thinking of molecules as having definite fixed geometries with the atoms vibrating about fixed angles and distances. Whenever a vibration, or other molecular motion, can carry the molecule from one structure to another at a detectable rate, the molecule is said to be **stereochemically nonrigid**.

1. Types
   a. Tautomerism or isomerism—if the two forms have different properties.
   b. Fluxional—if the two forms are chemically equivalent.
2. Example—PF_5
   a. X-ray data show two distinct sets of Fs. Two P—F bonds of length 1.577 Å (axial Fs) and three P—F bonds of length 1.534 Å (equatorial Fs).
   b. Would expect two different F resonances in the ^{19}F NMR. However, only a single resonance is observed.
   c. X-ray, electronic, and vibrational spectroscopy have interaction times on the order of 10^{-18} to 10^{-15} s; NMR interaction times are larger at 10^{-1} to 10^{-9} s. Therefore, a rate process in the order of 10^1 to 10^9 s^{-1} can be detected by NMR.

3.2 **Specific examples**

1. Three-coordinate compounds can exhibit inversion of configuration.
   a. The inversion of NH_3 and the plot of activation energy vs. reaction coordinate for the inversion process are given in Fig. 2.20.

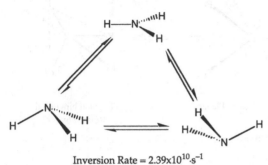

Inversion Rate = $2.39 \times 10^{10} \cdot s^{-1}$

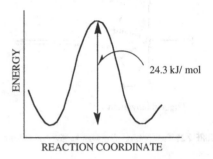

■ **FIGURE 2.20** Inversion of NH_3 and plot of activation energy versus reaction coordinate.

   b. Other $MR_1R_2R_3$ compounds.
      i. M = N, barriers are of the order 29–41 kJ/mol; have classical transitions of the energy barriers. The $NR_1R_2R_3$ compounds are, in principle, chiral, but they cannot be resolved due to a high degree of fluxionality.
      ii. M = P, E_as (barrier heights in Fig. 2.20) > 83 kJ/mol; M = As, E_as > 250 kJ/mol. Therefore, these phosphines can be resolved.
2. Four-Coordinate Compounds.
   There are no examples of stereochemical nonrigidity in tetrahedral complexes; the barriers are too large. It has been estimated that the rate of permeation or scrambling of the hydrogens in $CH_4 \sim 10^{-12} \, yr^{-1}$.
3. Five-Coordinate Compounds
   Stereochemical nonrigidity is a common facet of five-coordinate compounds.

a. Berry Mechanism (pseudorotation mechanism).

Mixing is thought to go through a square planar intermediate or transition state (Fig. 2.21). Note that one group remains in the trigonal plane.

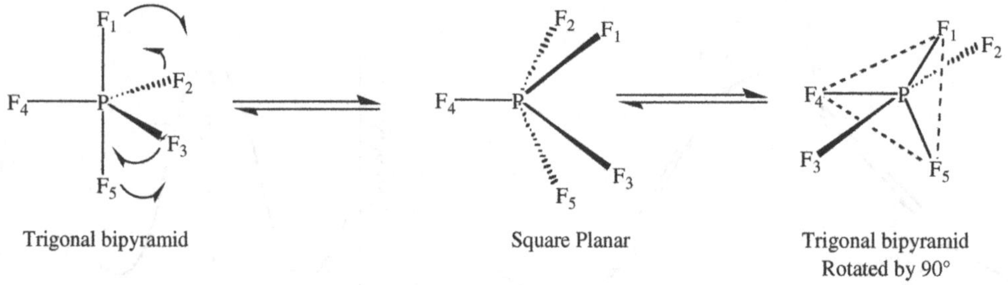

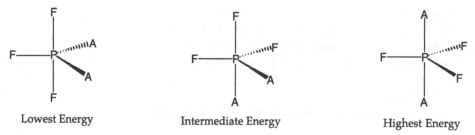

Trigonal bipyramid   Square Planar   Trigonal bipyramid
                                                                       Rotated by 90°

■ **FIGURE 2.21** Pseudorotation mechanism for distortion of five-coordinate compounds.

b. General type of mixing by going through idealized polyhedra = **polytopal isomerization**.

c. Fluxionality in compounds with different groups depends on the relative stabilities of the various isomers. More electronegative groups prefer to be in the axial position (Bent's rule = more electronegative groups prefer to bond to orbitals with less s character).

    i. APF_4 Fluxional since one can scramble the F's while A remains in an equatorial position.

    ii. A_2PF_3 Depends on the electronegativity of A relative to a fluorine atom (Fig. 2.22).

Lowest Energy   Intermediate Energy   Highest Energy

■ **FIGURE 2.22** Energy-dependent configurations of A_2PF_3 molecule.

$(R_3P)_2PF_3$ is stereochemically rigid as found in X-ray structure and in ^{19}F NMR spectrum. The ^{19}F NMR spectrum of Cl_2PF_3 shows equivalent F's above $-60°C$, while nonequivalent F's are found below $-60°C$.

4. Six-Coordinate Compounds.

Most octahedral complexes are stereochemically rigid. However, there are some cases of intramolecular isomerization.

   a. $(R_3P)_4RuH_2$ and $(R_3P)_4FeH_2$ show intramolecular *cis–trans* isomerism.

   b. Some six-coordinate *tris*(bidentate) complexes, which are optically active, undergo intramolecular racemization. Example $[Co(en)_3]^{3+}$ (en $= NH_2CH_2CH_2NH_2$) (Fig. 2.23).

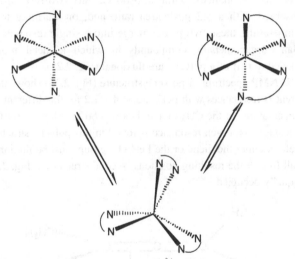

■ **FIGURE 2.23** Intramolecular racemization in $[Co(en)_3]^{3+}$ molecule.

5. Seven-Coordinate Compounds.

   a. Most seven-coordinate complexes show ligand equivalency by NMR and are presumed to be fluxional.

   b. Examples: both IF_7 and RuF_7 show only one resonance in the ^{19}F NMR spectroscopy.

6. Organometallic Compounds.

a. The organometallic iron carbonyl compound has a Cs symmetry (described in Chapter 3) with four different COs (Fig. 2.24).

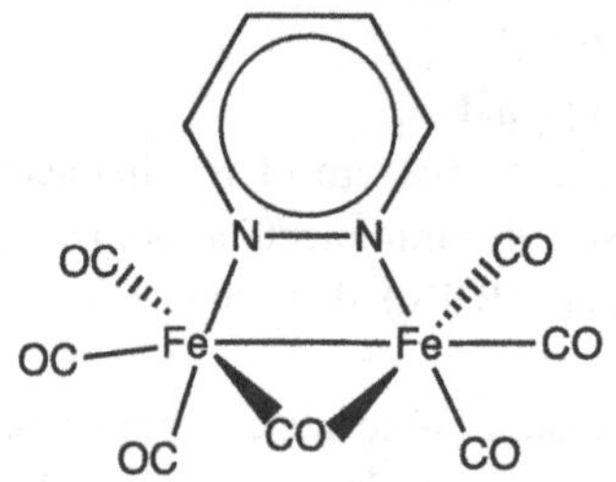

■ **FIGURE 2.24** Molecular structure of metal carbonyl [$Fe_2(CO)_7(C_4H_4N_2)$].

At and below $-139°C$, the ^{13}C NMR spectra show all of the predicted resonances, while at $-60°C$, only two resonances are observed with a 5:2 peak area ratio and, on warming to room temperature, these two peaks merge into a single resonance.

b. Many organometallic compounds in which a metal moiety is bonded to part of a polyene are fluxional (Fig. 2.25).

The ^1H NMR spectrum of polyene structure (Fig. 2.25) shows three different H resonances with peak areas of 1:1:2 for the different hydrogen atoms of the CH_3 moiety below $-60°C$. Above $-60°C$, only a single hydrogen resonance is found in this polyene structure. Therefore, either the diene or the $Fe(CO)_3$ group must be fluxional so that all four of the resulting positions (second structure in Fig. 2.25) are equally occupied.

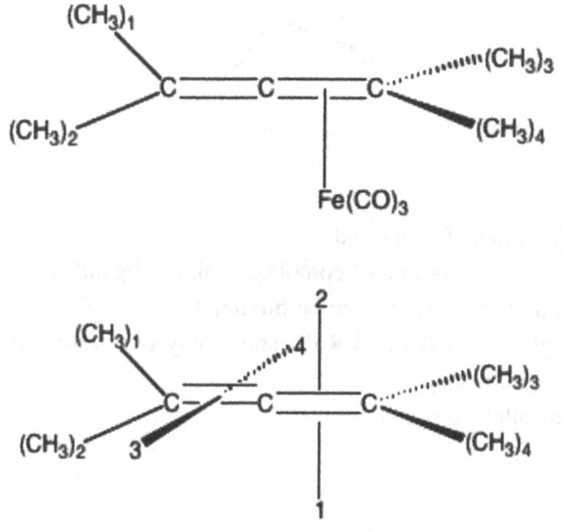

■ **FIGURE 2.25** Fluxional molecule of organoiron carbonyl [π—$Fe(CO)_3C_3(CH_3)_4$].

Molecular Symmetry—Part I: Point Group Assignment

1. INTRODUCTION: IS IT NECESSARY TO LEARN MOLECULAR SYMMETRY?

Nature created shapes, and humans have classified them into groups or patterns. Whether in biology, mathematics, arts, engineering, physics, or chemistry, these classifications were created to simplify the objects or the world around us. Similarly, chemists classify molecules based on their symmetry and the collections of symmetry elements, such as points, lines, and planes that intersect at a single point, to form a "Point Group". Obviously, in nature's cycle of events, *geometry* of a molecule points to a *symmetry group* leading to *group representation* that determines the *structural properties* based on the *geometry* of that particular species. Thus, for any molecule's symmetry, one can predict important properties leading to molecular identity, such as space group of any crystalline form, chirality (optical activity) or lack of chirality, overall polarity (dipole moments), infrared spectrum, and Raman spectrum. One must admit that symmetry is the consistency, that is, the repetition, of an object in space and/or in time as we normally observe in a wall drawing/painting, wings of a butterfly, flower petals, musical notes, and even the repetition of day and night and the seasons on our planet. Since symmetry is an important aspect of nature, we must learn about its mystery.

2. ELEMENTS OF SYMMETRY

2.1 Symmetry operations

1. The shape of a molecule is described by indicating the spatial arrangement of the atoms.
    a. For simple symmetric molecules, the terms trigonal planar, tetrahedral, octahedral, etc. are useful descriptions. However, for more

Advanced Inorganic Chemistry. http://dx.doi.org/10.1016/B978-0-12-801982-5.00003-5

complex molecules not having a great deal of symmetry, a better way of describing the stereochemical arrangement of the atoms is needed.

   b. Example: Consider the pentagonal bipyramidal molecule PCl_3F_2 (Fig. 3.1). There are three ways to arrange the atoms.

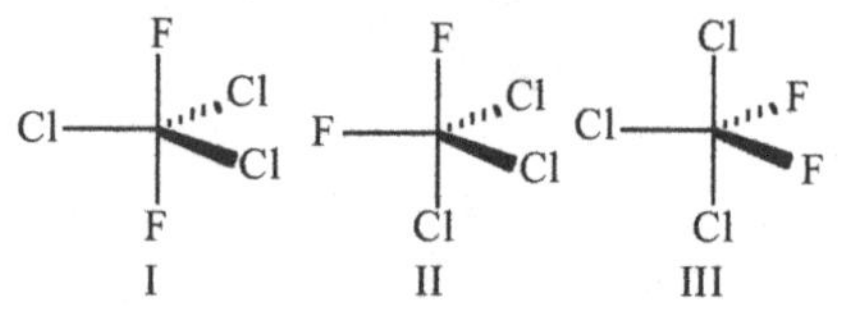

■ **FIGURE 3.1** Possible isomers of PCl_3F_2.

   c. All three structures have a trigonal bipyramidal structure. Of the three, I is the most symmetric, whereas II is the least symmetric. This ordering is done on the basis that in I, all three Cl's are equivalent, as are the two F's. In III, two Cl's and two F's are equivalent, whereas in II, only two Cl's are equivalent. Therefore, one can use symmetry to describe molecular shapes.

2. Only four symmetry operations are needed to define a structure. These consist of rotations, reflections, inversion about a point, and rotation and reflection in a perpendicular mirror plane.

3. If one carries out a symmetry operation and obtains an equivalent (indistinguishable) structure, the molecule is said to possess that particular element of symmetry.

2.2 **Operations and elements**

1. Rotation about an axis.

   a. The symbol C_n will stand for "n" number of rotations by an angle of 360 degrees ($2\pi/n$).

      i. C_1 = rotation by 2π (360 degrees). This is a trivial operation.
           C_2 = rotation by $2\pi/2$ (180 degrees).
           C_3 = rotation by $2\pi/3$ (120 degrees).
           C_4 = rotation by $2\pi/4$ (90 degrees).

      ii. A single rotation by $2\pi/n = C_n$.
           A rotation by $2(2\pi/n)$ [carry out the operation twice] $= C_n^2$.
           A rotation by $3(2\pi/n)$ [carry out the operation three times] $=$ C_n^3.

      iii. The various C_4 operations are shown for the vertices of a regular octahedron (Fig. 3.2).

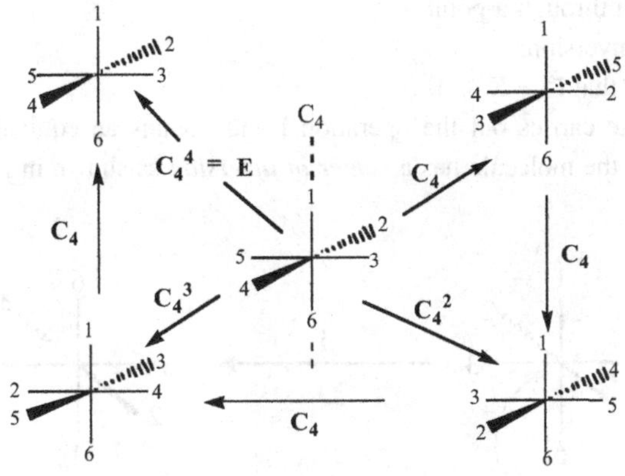

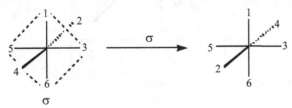

■ **FIGURE 3.2** Illustration of rotation about an axis.

iv. If one applies C_n n times (C_n^n), it will have a total rotation of 2π (360 degrees).

This will give not an *equivalent* structure, but an *identical* structure.

This operation is called the identify operation and has the symbol **E**.

b. If one carries out the operation C_n and obtains an equivalent structure, the molecule has an *n-fold proper axis of symmetry*. The symbol for this element of symmetry is C_n. Since the symmetry element and the operation have the same symbol, we will use bold lettering to indicate the operation.

c. The element C_n generates n operations $[C_n, C_n^2, C_n^3, \ldots C_n^n(=E)]$.

2. Reflection in a mirror plane.

a. σ = reflection in a plane.

σ^2 = reflection twice = **E**.

b. If one carries out the operation σ and obtains an equivalent structure, the molecule is said to possess a plane of symmetry. The symbol for this element is σ. An example is shown in Fig. 3.3 (dashed lines connect the atoms in the mirror plane).

■ **FIGURE 3.3** Illustration of reflection in a mirror plane.

c. Each element σ generates only one unique operation since $\sigma^2 = E$.

3. Inversion through a point.

    a. i = inversion.

    Note that $i^2 = E$.

    b. If one carries out the operation **i** and obtains an equivalent structure, the molecule has a *center of inversion* as shown in Fig. 3.4.

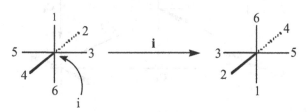

■ **FIGURE 3.4** Illustration of center of inversion.

The symbol for this element is *i*.

4. Rotation + reflection in a perpendicular mirror plane.

    a. S_n = rotation by $2\pi/n$ followed by reflection in a plane perpendicular to the axis of rotation. Such a plane is called a horizontal plane (σ_h).

    b. If one carries out this operation and obtains an equivalent structure, the molecule has an *n-fold improper axis of rotation* (Fig. 3.5).

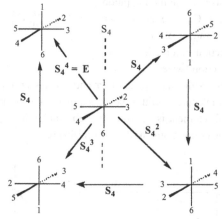

■ **FIGURE 3.5** Illustration of improper axes of rotation.

The symbol for this element is S_n.

    c. The element S_n gives rise to n operations (S_n, S_n^2 S_n^3, S_n^4, ... S_n^n).

2.2.1 **Examples**

Example 1. Symmetry elements and operations for PBr₅.

PBr₅ is a trigonal bipyramidal molecule with two axial and three equatorial Br's (Fig. 3.6). We will number the positions to see the results of symmetry operations.

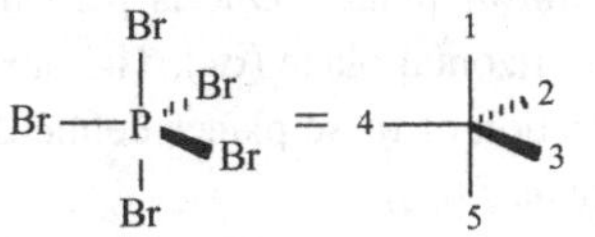

■ **FIGURE 3.6** Atom numbering surrounding the central phosphorus.

Note the axial positions are 1 and 5, whereas the equatorial positions are 2, 3, 4.

1. Proper axes of rotation
   There is one C_3 and three C_2 axes. These axes and their related operations are shown in Fig. 3.7.

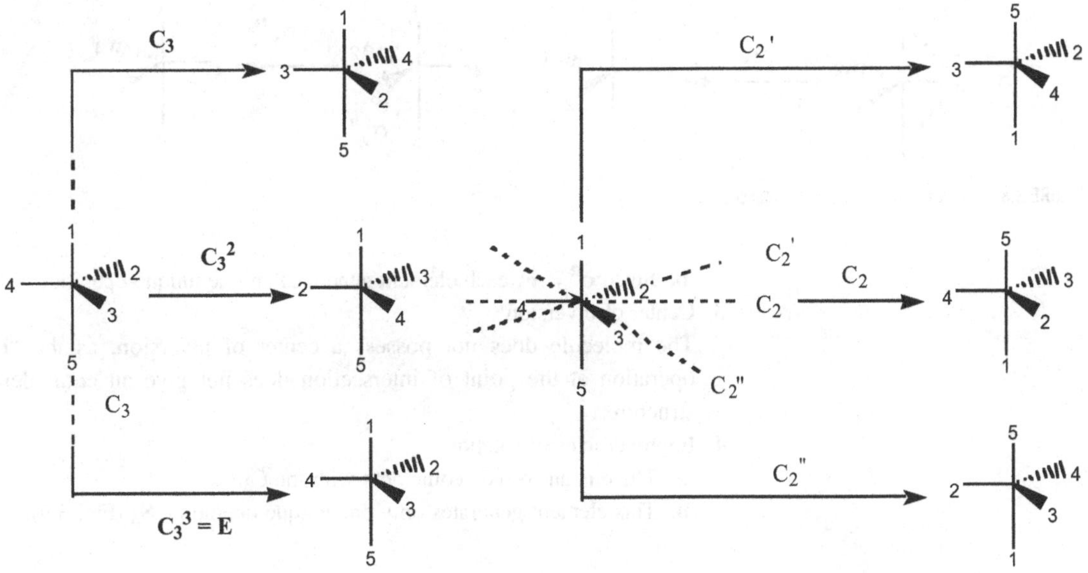

■ **FIGURE 3.7** Illustration of proper axes of rotation.

The proper axis of rotation with the highest symmetry is taken as the main symmetry element (for PBr_5 it is the C_3 axis). All other symmetry elements will be referred to this element. Note that each C_2 axis generates only one unique operation since $C_2^2 = E$, and E was already listed as C_3^3. The four symmetry elements generate six unique symmetry operations.

2. Mirror planes

   a. There are four mirror planes. One is perpendicular to the C_3 axis and is called a horizontal plane (σ_h). The other three contain the C_3 axis. The intersection of these planes define the C_3 axis; they are called vertical planes (σ_v).

   These planes and their operations are shown in Fig. 3.8. A dashed line is used to connect the atoms in the particular plane.

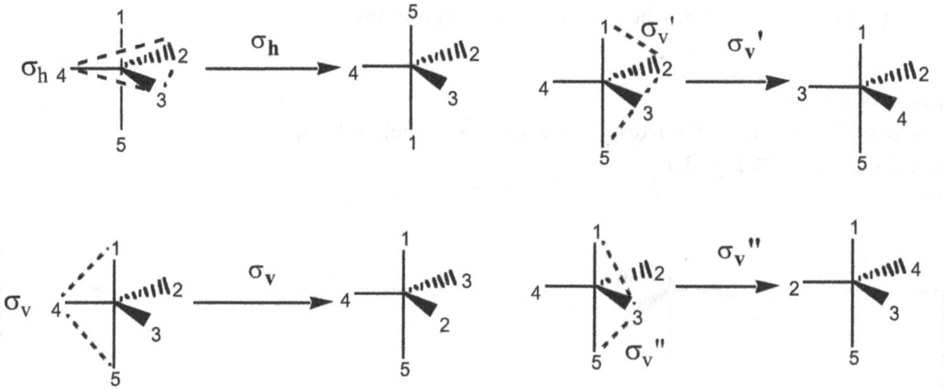

FIGURE 3.8 Illustration of mirror plane of symmetry.

   b. Since $\sigma^2 = E$, each element generates on one unique operation.

3. Center of inversion

   The molecule does not possess a center of inversion, as the "i" operation at the point of intersection does not give an equivalent structure.

4. Improper axes of rotation

   a. There is an S_3 axis coincident with the C_3.

   b. This element generates only one unique operation, S_3 (Fig. 3.9).

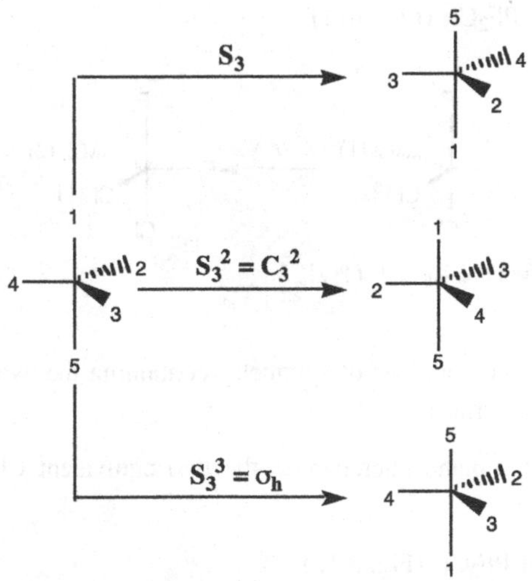

■ **FIGURE 3.9** Operation of improper axes of rotation.

5. Summary

The molecule has nine symmetry elements (C_3, $3C_2$, $3\sigma_v$, σ_h, S_3) that generate a total of 11 unique symmetry operations.

Example 2. The elements and operators for the isomers of PF_2Cl_3 (Fig. 3.10).

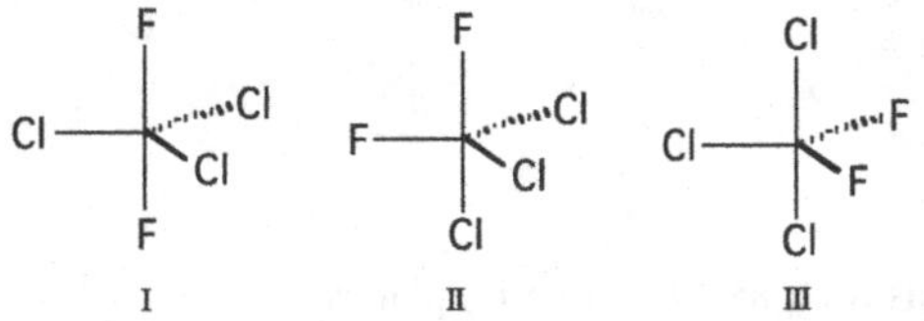

■ **FIGURE 3.10** Isomers of PF_2Cl_3.

1. Isomer I

This isomer has all of the symmetry elements and operations of PBr_5 outlined earlier.

In PBr_5, there were two sets of Br's: the two axial atoms and the three equatorial ones. No symmetry operation can interchange the axials with equatorials. Therefore, nothing changes if the axial groups and the equatorial groups are different.

2. Isomer II of PF_2Cl_3 (Fig. 3.11)

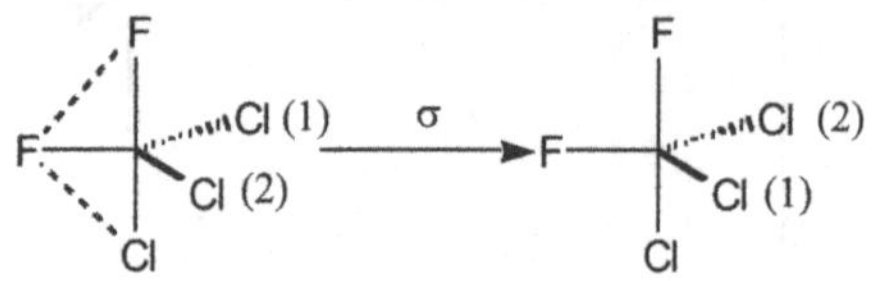

■ **FIGURE 3.11** Reflection in isomer II of PF_2Cl_3.

This isomer has only a plane of symmetry containing the two axial groups and the one equatorial F.

Reflection in this plane interchanges the two equivalent Cl's, Cl(1) and Cl(2).

3. Isomer III of PF_2Cl_3 (Fig. 3.12)

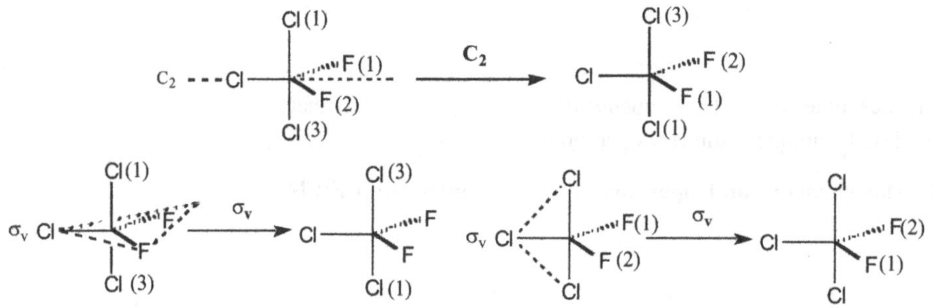

■ **FIGURE 3.12** Rotation and reflection in isomer III of PF_2Cl_3.

This isomer has a C_2 axis and two σ_v planes.

3. **POINT GROUPS**

3.1 **Introduction**

1. All symmetry elements intersect at a point, the center of symmetry of the molecule.

   a. This point has all the symmetry of the molecule.

   b. On the basis of symmetry, one can classify molecules into *point groups*.

2. Molecules with the same symmetry are in the same point group. We will describe the shape of the molecule by assigning it to a particular point group.

3.2 **Rules for assigning point groups**

1. Look for proper axes of symmetry. If none exists and there is only:
   a. a plane of symmetry, then the point group is C_s.
   b. a center of inversion, then the point group is C_i.
2. If proper axes exist, pick the highest symmetry axis as the major axis. If no uniquely high axis exists, pick one with some special geometric property, e.g., being colinear with some unique bond, as the major axis. If there is no obviously unique axis, just pick one and stick with it.
3. Once the major C_n is found, look for an S_{2n} axis coincident with it. If one is found and no other element of symmetry exists (other than i), the point group is S_n (n will always be even).
4. If no S_{2n} exists or if other symmetry elements exist, then:
   a. look for a set of n C_2 axes perpendicular to C_n. If such a set exists, the point group is D_n, D_{nh}, or D_{nd}.
   b. if a set of n C_2 axes does not exist, then the point group is C_n, C_{nv}, or C_{nh}.
5. If the molecule is one with a D point group:
   a. look for a mirror plane perpendicular to the major C_n (this is called a horizontal plane and has the symbol σ_h). If this plane exists, the point group is D_{nh}.
   b. if no σ_h plane exists, then look for a set of planes bisecting the C_2 axes (these planes are called dihedral planes and have the symbol σ_d). If these planes are found, the point group is D_{nd}.
   c. if neither σ_h nor σ_d planes are found, the point group is D_n.
6. If the molecule is one with a C point group:
   a. look for a σ_h plane. If one is found the point group is C_{nh}.
   b. look for a set of n σ_v planes (planes vertical to and containing the C_n axis).
   If a set is found, the point group is C_{nv}.
   c. if neither σ_h nor σ_v planes exist, the point group is C_n.
7. Special groups (these should be considered first):
   a. $C_{\infty v}$: Linear molecules that do not possess a plane of symmetry perpendicular to the C_∞ axis (example is HCl).
   b. $D_{\infty h}$: Linear molecule possessing a plane of symmetry perpendicular to the C_∞ axis (example is CO_2 or Cl_2).
   c. T_d: Tetrahedral molecules (example is CH_4).
   d. O_h: Octahedral molecules (example is SF_6).

3.3 **Examples**

1. PBr_5 and PCl_3F_2 are D_{3h} (have $C_3 + 3C_2$'s $+ \sigma_h$ symmetry elements).
2. NH_3 is C_{3v}.
3. H_2O is C_{2v}.
4. Unfavorable isomers of PCl_3F_2 (Fig. 3.13).

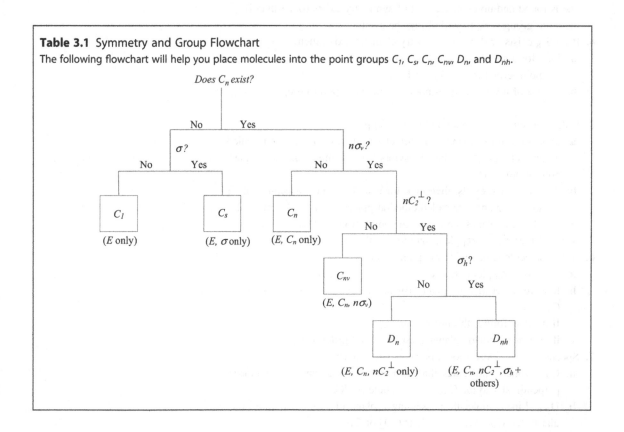

■ **FIGURE 3.13** Isomers of PCl_3F_2.

A selection of a number of molecules and their point groups determined with the accompanying flowchart (Table 3.1; reproduced from the book

Table 3.1 Symmetry and Group Flowchart

The following flowchart will help you place molecules into the point groups C_1, C_s, C_n, C_{nv}, D_n, and D_{nh}.

Does C_n exist?

```
                                No        Yes
                          ┌──────────┴──────────┐
                        σ?                     nσ_v?
                    ┌────┴────┐            ┌─────┴─────┐
                   No        Yes          No          Yes
                                                       │
                                                      nC_2^⊥?
              ┌─────┐   ┌─────┐   ┌─────┐        ┌──────┴──────┐
              │ C_1 │   │ C_s │   │ C_n │       No            Yes
              └─────┘   └─────┘   └─────┘                      │
                                                              σ_h?
            (E only)  (E, σ only) (E, C_n only)        ┌───────┴───────┐
                            ┌─────┐                    No             Yes
                            │ C_nv│
                            └─────┘
                          (E, C_n, nσ_v)       ┌─────┐          ┌──────┐
                                               │ D_n │          │ D_nh │
                                               └─────┘          └──────┘
                                        (E, C_n, nC_2^⊥ only) (E, C_n, nC_2^⊥, σ_h +
                                                                    others)
```

An Introduction to Inorganic Chemistry, K. F. Purcell and J. C. Kotz; Saunders College Publishing, Philadelphia, **1980**. 637 pp) as well as the structures are shown in Table 3.2 [reproduced from the Website funded under

Table 3.2 Symmetry and Point Group

Number of Stereochemically Active e⁻ Pairs	Example	Point Group
2	$O{=}C{=}O$	$D_{\infty h}$
2	$H{-}C{\equiv}N$	$C_{\infty v}$
3		D_{3h}
3		C_{2v}
3		C_S
4		T_d
4		C_{3v}
4		C_{2v}

Continued

Table 3.2 Symmetry and Point Group *continued*

Number of Stereochemically Active e⁻ Pairs	Example	Point Group
5		D_{3h}
5		C_{3v}
5		C_{2v}
6		O_h
6		D_{4h}
6		C_{4v}

NSF-DUE # 0536710 (copyrighted in 2014 by Dean H. Johnston and Otterbein University (http://symmetry.otterbein.edu/challenge/index.html))].

Chapter 4

Group Theory: Matrix Representation and Character Tables

1. INTRODUCTION: IS IT NECESSARY TO LEARN GROUP THEORY?

Symmetry was first invoked by theoreticians in taking representation theory from point groups to crystal structures of solids. According to quantum mechanics, electrons and protons are truly identical with exact symmetry operations of rotation and reflection in a subatomic level whose properties are exposed by group theory, which is based on mathematical interpretation. Group theory, also known as the theory of representations in physics, utilizes matrices with certain members of a symmetrical vector space that can be classified by their symmetry. Consequently, the observed spectroscopic states of atoms and molecules correspond to symmetrical functions that lead to **selection rules**, which, in turn, indicate which transitions are allowed and/or observed in spectroscopy. Thus, group theory is as important as a map and it gives us a lot more information than you might imagine.

2. OTHER PROPERTIES OF SYMMETRY OPERATIONS

2.1 Sequential operations

Consider the NH_3 molecule in the C_{3v} point group.

1. The symmetry operations are: E, C_3, C_3^2, σ_v, σ_v', σ_v''. Fig. 4.1 represents a view of the NH_3 molecule from above the N center and the results of carrying out some symmetry operations.

Advanced Inorganic Chemistry. http://dx.doi.org/10.1016/B978-0-12-801982-5.00004-7

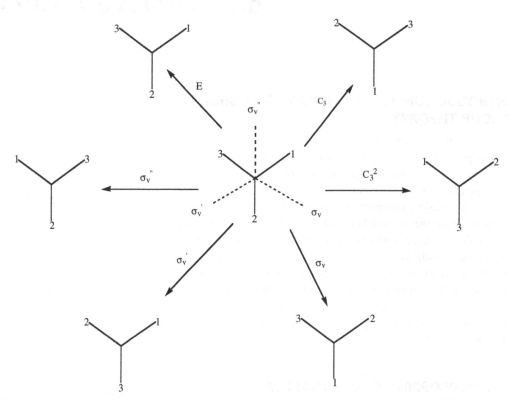

■ **FIGURE 4.1** Symmetry operations in the NH_3 molecule.

2. Now consider the results of carrying out several symmetry operations sequentially.

    a. C_3 followed by a σ_v ($\sigma_v C_3$), this is the same as a σ'_v (Fig. 4.2).
       We have the multiplication equation $\sigma_v C_3 = \sigma'_v$.

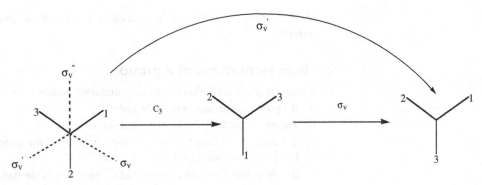

■ **FIGURE 4.2** Sequential symmetry operations in the NH₃ molecule.

b. In the same way, one can see: $C_3 C_3^2 = E$; $\sigma_v \sigma_v = E$; $C_3^2 C_3^2 = C_3$.

Note that the order is important. Fig. 4.3 shows the result when the sequence is $C_3 \sigma_v (= \sigma_v'')$.

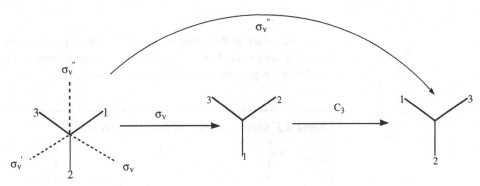

■ **FIGURE 4.3** Symmetry operations, $C_3 \sigma_v (= \sigma_v')$, in the NH₃ molecule.

3. The multiplication table, shown in Table 4.1, can be constructed. The order is (Row) × (Column) = Product.

Table 4.1 Multiplication Order of Symmetry

C_{3v}	E	C_3	C_3^2	σ_v	σ_v'	σ_v''
E	E	C_3	C_3^2	σ_v	σ_v'	σ_v''
C_3	C_3	C_3^2	E	σ_v''	σ_v	σ_v'
C_3^2	C_3^2	E	C_3	σ_v'	σ_v''	σ_v
σ_v	σ_v	σ_v'	σ_v''	E	C_3	C_3^2
σ_v'	σ_v'	σ_v''	σ_v	C_3^2	E	C_3
σ_v''	σ_v''	σ_v	σ_v'	C_3	C_3^2	E

These symmetry operations constitute a mathematical **group** of **order 6**.

2.2 **Representations of a group**

1. A group is just a collection of elements that obey certain rules.
   a. All products are members of the group.
   b. The group contains the identity element, E.
   c. The reciprocal of each element is also a member of the group.
      i. A^{-1} = reciprocal of A if $A^{-1}A = AA^{-1} = E$
      ii. Note that C_3 is the reciprocal of C_3^2 and the σ's are their own reciprocals.
   d. The associative law of multiplication holds. That is:

   $$\sigma_v(C_3\sigma'_v) = (\sigma_v C_3)\sigma'_v$$

   $$\sigma_v\sigma_v = \sigma'_v\sigma'_v$$

   $$E = E$$

   e. The **order of the group** = number of elements in the group.
2. Any collection of elements that obey the aforementioned multiplication table is a **representation** of this group (Table 4.2).

Table 4.2 Some Examples of Group Representation

Γ_1	E	C_3	C_3^2	σ_v	σ'_v	σ''_v
Γ_2	1	1	1	1	1	1
Γ_3	1	1	1	-1	-1	-1
Γ_4	E	A	B	C	D	F

3. Matrix representations.
   a. A set of matrices that will transform the x, y, z coordinates of each atom in the NH_3 molecule into their new coordinates on each symmetry operation in the C_{3v} point group is also a representation of the group.
   b. Orient the NH_3 molecule so that the N is on the z axis and the H's are arranged parallel to the xy plane as shown in Fig. 4.4.

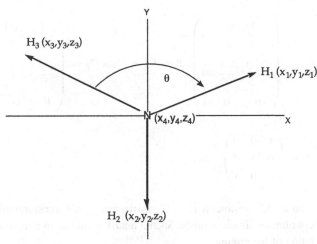

■ **FIGURE 4.4** Orientation of the NH_3 molecule with the center on the z-axis.

c. Carry out a σ_v reflection in the yz plane and see how the coordinates of H_1 change.

i. Instead of being x_1, y_1, and z_1, they are now $-x_1$, y_1, and z_1.

ii. The following matrix can be constructed to do this:

$$\sigma \begin{pmatrix} x_1 \\ y_1 \\ z_1 \end{pmatrix} = \begin{pmatrix} -1 & 0 & 0 \\ 0 & 1 & 0 \\ 0 & 0 & 1 \end{pmatrix} \begin{pmatrix} x_1 \\ y_1 \\ z_1 \end{pmatrix} = \begin{pmatrix} -x_1 \\ y_1 \\ z_1 \end{pmatrix}$$

so that

$$\sigma = \begin{pmatrix} -1 & 0 & 0 \\ 0 & 1 & 0 \\ 0 & 0 & 1 \end{pmatrix}$$

The coordinates are said to form a **basis** for the group.

iii. In the same way, matrices for the other symmetry transformations can be derived. Some are:

$$C_\theta = \begin{pmatrix} \cos\theta & \sin\theta & 0 \\ -\sin\theta & \cos\theta & 0 \\ 0 & 0 & 1 \end{pmatrix}; \text{ for } C_3, \theta = 120° = \frac{2\pi}{3}$$

$$C_3 = \begin{pmatrix} -\dfrac{1}{2} & \dfrac{\sqrt{3}}{2} & 0 \\ -\dfrac{\sqrt{3}}{2} & -\dfrac{1}{2} & 0 \\ 0 & 0 & 1 \end{pmatrix}, \text{ then } C_3^2 = C_3 \times C_3 = \begin{pmatrix} -\dfrac{1}{2} & -\dfrac{\sqrt{3}}{2} & 0 \\ \dfrac{\sqrt{3}}{2} & -\dfrac{1}{2} & 0 \\ 0 & 0 & 1 \end{pmatrix}$$

$$E = \begin{pmatrix} 1 & 0 & 0 \\ 0 & 1 & 0 \\ 0 & 0 & 1 \end{pmatrix} \quad \text{etc.}$$

d. You could construct a 12×12 matrix that would transform all the coordinates simultaneously. Such a matrix would form a representation of the group.

It would consist of matrices of the type mentioned earlier along the diagonal with zeros elsewhere.

2.3 Reducible and irreducible representations

1. It is possible to simplify large matrices that form representations of a group into simpler ones consisting of lower dimensional matrices along the diagonal and zeros elsewhere.

   a. The large matrices are **reducible representations** of the group.

   b. Each of the smaller matrices along the diagonal also forms a representation of the group.

   c. If you repeat the simplification process until the smaller matrices cannot be simplified any more, they are called **irreducible representations** of the group.

   d. Example: The C_θ matrix is a 3×3 reducible representation if the group is composed of the irreducible representation

   $\begin{pmatrix} \cos\theta & \sin\theta \\ -\sin\theta & \cos\theta \end{pmatrix}$ and (1). That is, the 3×3 matrix contains a

   2×2 and a 1×1 irreducible representation.

   e. Any reducible representation of a point group can be resolved into a limited number of irreducible ones.

      i. The irreducible representations for each point group have been worked out.

      ii. The irreducible representations have properties that are useful in molecular structural analyses.

2. Characters of matrices.

    a. For molecular point groups, all the matrices will be square matrices, that is, the number of rows equal the number of columns.

    b. The **character** of a square matrix, χ = sum of the elements along the diagonal.

    c. It turns out that the useful properties of the irreducible representations are also reflected in their characters. Therefore we do not have to deal with the whole matrices, but only their characters.

    d. The characters of the irreducible representations have been worked out for the molecular point groups and are listed in **character tables**.

3. For a molecule containing N atoms and belonging to a particular point group, a series of $3N \times 3N$ matrices can be constructed for each symmetry operation that correctly transforms the coordinates of each atom in the molecule.

    a. In general, these will be reducible representations of the group. The irreducible ones must be determined.

    b. Let R = a symmetry operation of the point group.

    c. $\chi_{(Red)}(R)$ = the character of the reducible representation for the operation R.

    d. $\chi_j(R)$ = the character for the jth irreducible representation for the same symmetry operation.

    e. a_j = the number of times that the jth irreducible representation appears in the reducible one.

4. It can be shown that:

$$a_j = \frac{1}{h} \sum_R \chi_{(Red)}(R)\chi_j(R)$$

where h = the order of the group = number of symmetry operations in the group.

Using this **reduction formula**, one can obtain the irreducible representations contained in a reducible one.

The sum, $\sum$, is over all the operations in the group.

    a. Example: C_{3v} point group.

        i. The point group is composed of three irreducible representations, Γ_1, Γ_2, and Γ_3.

        ii. The following character table (Table 4.3) lists the characters of these representations as well as those of a reducible representation Γ_R.

Table 4.3 Representations for a C_{3v} Point Group

	E	C_3	C_3^2	σ_v	σ_v'	σ_v''
Γ_1	1	1	1	1	1	1
Γ_2	1	1	1	−1	−1	−1
Γ_3	2	−1	−1	0	0	0
Γ_R	5	2	2	−1	−1	−1

Calculate the number of each of the irreducible representations contained in Γ_R:

$$a_j = \frac{1}{h} \sum_R \chi_{(Red)}(R)\chi_j(R)$$

$$a_1 = \frac{1}{6}[5x1 + 2x1 + 2x1 + (-1)x1 + (-1)x1 + (-1)x1] = 1$$

$$a_2 = \frac{1}{6}[5x1 + 2x1 + 2x1 + (-1)x(-1) + (-1)x(-1) + (-1)x(-1)] = 2$$

$$a_3 = \frac{1}{6}[5x2 + 2x(-1) + 2x(-1) + (-1)x0 + (-1)x0 + (-1)x0] = 1$$

These calculations result in $\Gamma_R = \Gamma_1 + 2\Gamma_2 + \Gamma_3$.

2.4 Character tables

Consider the character table (Table 4.4) for the C_{3v} point group.

Table 4.4 A Typical Character Table for the C_{3v} Point Group

C_{3v}	E	$2C_3$	$3\sigma_v$		
A_1	1	1	1	z	$x^2 + y^2, z^2$
A_2	1	1	−1	R_z	
E	2	−1	0	(x, y) (R_x, R_y)	$(x^2 - y^2, xy)$ $(xz - yz)$
I	II			III	IV

1. Region I:
   a. Point group at the top.
   b. Symbols for the irreducible representations
      i. A, B = one-dimensional irreducible representations.
         E = two-dimensional irreducible representation.
         T = three-dimensional irreducible representations.

ii.   The notations tell something about the symmetry of the irreducible representation. The *B* designation indicates the character of the irreducible representation, C_n, is -1, with an *A* indicating a $+1$ character. In addition to the letters, the symbols may carry (′) or (″) notations or have numerical and/or letter subscripts. For example, in the C_{3v} point group, the subscript 2 in A_2 indicates that the irreducible representations in the σ_v class have characters of -1. In point groups where there is a center of inversion, the irreducible representations will also be labeled with subscripts *g* (*gerade* = does not change sign on inversion) or *u* (*ungerade* = changes sign on inversion).

2. Region II:
   a. Symmetry operations and their characters.
   b. Symmetry operations are grouped in **classes**. Elements in the same class have the same χ's. $2C_3 = C_3$ and C_3^2; $3\sigma_v = \sigma_v, \sigma_v', \sigma_v''$.

3. Region III:
   a. *x*, *y*, and *z* either alone or in groups. These are the bases for the various irreducible representations.
   b. R_x, R_y, and R_z either alone or in groups represent rotations about the particular axes.

4. Region IV:
   a. The squares and binary products that also serve as bases for the particular irreducible representations.

3. APPLICATIONS TO MOLECULAR STRUCTURE AND PROPERTIES

3.1 Application to quantum mechanics

1. In molecular orbital theory, we construct the molecular orbitals (ψ_{MO}'s) in a molecule as a linear combination of atomic orbitals (ϕ_I) of the involved atoms. That is, $\psi_{MO} = \sum c_I \, \phi_i$. The process of selecting the correct weighting coefficients, c_I, can be quite lengthy and tedious. However, we can take advantage of the symmetry of the molecule to greatly simplify the process.

2. A wave function, ψ, can describe either a degenerate or a nondegenerate energy state (level).

3. When you carry out a symmetry operation on a molecule and obtain an equivalent structure, you also carry out the operation on the wave function. Therefore, the ψ's must form a basis for the representation of the group.

a. If R is a symmetry operation in the group, then
$R|\psi|^2 = |\psi|^2$. The electron density cannot change.
$H(R\psi) = E(R\psi)$. The energy cannot change.

b. If ψ is a nondegenerate energy state, then $R\psi = \pm\psi$. That is, at most, the wave function can only change sign. Therefore, ψ must form the basis of a one-dimensional irreducible representation of the point group.

c. If ψ is a member of an n-fold degenerate energy state, then at most, R will transform ψ into a linear combination of the wave functions in that state. Therefore, the wave functions in an n-fold degenerate energy state form a basis from a representation of the group. This also is an irreducible representation.

4. In quantum mechanics, one encounters integrals of the form $\int_{\substack{all \\ space}} \phi_i^* P \phi_j d\tau$, where P is an operator and the ϕ's are wave functions or approximate wave functions. It can be shown that these integrals are zero unless the function $\phi_i^* P \phi_j$ forms a basis for a representation of the group that contains a totally symmetric irreducible representation. That is, a one-dimensional irreducible representation where the characters are all $+1$.

a. Suppose that ϕ_1 forms the basis of an irreducible representation, Γ_1, and ϕ_2 forms a basis for the irreducible representation, Γ_2. The function $\phi_1\phi_2$ forms a basis of a representation of the group, Γ_3, that is, the **direct product** of Γ_1 and Γ_2.
$(\Gamma_3 = \Gamma_1 \times \Gamma_2)$ Γ_3 is not necessarily an irreducible representation.

b. It can be shown that the characters of Γ_3 for a particular symmetry operation, R, is equal to the product of the characters of Γ_1 and Γ_2 for that operation. That is:

$$\chi_{\Gamma_3}(R) = \chi_{\Gamma_1}(R)\chi_{\Gamma_2}(R)$$

By using the formula $a_j = \frac{1}{h}\sum_R \chi_{(Red)}(R)\chi_j(R)$ it is possible to determine the irreducible representations that are contained in Γ_3.

c. In molecular orbital theory, one finds integrals of the form $\int_{\substack{all \\ space}} \phi_i^* H \phi_j d\tau$. If one could take a preliminary linear combination of the atomic orbitals, ϕ, to give **group symmetry orbitals** [Symmetry Adapted Linear Combinations (SALC)], which form the bases of the irreducible representations of the point group of the molecule, many of the integrals could be immediately set equal to zero. In this way, a considerable simplification could be achieved.

d. Since H must be totally symmetric to any symmetry operation in the group, the representation for which $\phi_1\phi_2$ forms a basis must contain the totally symmetric irreducible representation. This is true only if ϕ_1 and ϕ_2 form bases for the **same** irreducible representation.

  e. Since σ and π bonds do not interact with each other, they can be
     considered separately in molecules where they both appear.
5. One can use group theory to construct these SALCs. The formula for
   the generation of the SALC, called a **projection formula**, will be pre-
   sented without proof.
   a. Suppose ϕ is an atomic orbital that is contained in a symmetry
      orbital that forms the bases of the jth irreducible representation of
      dimension l_j that occurs a_j times. The SALC orbitals, Ψ_{kt}^j, can be
      generated by:

$$\sum_R \chi_j(R)R\phi = \frac{h}{l_j}\sum_{k=1}^{a_j}\sum_{t=1}^{l_j} c_{kt}\psi_{kt}^j.$$

   Although this looks complex, it can be readily interpreted.
   Consider a few examples.
   b. Suppose you have a one-dimensional irreducible representation that
      appears one time, then $\sum_R \chi_j(R)R\phi = hc_{11}^j\psi_{11}^j$ and ψ_{11}^j is a molec-
      ular orbital. That is, if you carry out a symmetry orbital on an
      atomic orbital, $R\phi$, this will yield another atomic orbital in the
      molecule. Thus, $\sum_R \chi_j(R)R\phi$ will generate an SALC of atomic or-
      bitals, which forms a basis for the particular irreducible representa-
      tion. If that irreducible representation occurs only once, then the
      SALC orbital is a molecular orbital.
   c. Suppose that one has a one-dimensional irreducible representation
      that occurs twice ($l_j = 1$, $a_j = 2$), then
      $\sum_R \chi_j(R)R\phi = hc_{11}{}^j\psi_{11}{}^j + hc_{21}^j\psi_{21}^j$. The molecular orbitals and
      their energies can be obtained by solving a 2×2 determinate. That
      is, $\psi_{MO} = c_1\psi_{11}^j + c_2\psi_{21}^j$.
   d. Suppose you have a two-dimensional irreducible representation
      that occurs one time ($l_j = 2$, $a_j = 1$). Then $\sum_R \chi_j(R)R\phi$ will yield
      two wave functions of the same energy or a linear combination of
      them that can be used as molecular orbitals.
   e. We will use the symbol $\hat{P}^j$ to stand for the projection operation.

3.2 Bonding in a triangular planar structure (AX₃ with D₃ₕ point group)

1. **σ bonding.** Since sigma bonds are directed along the bonding axes,
   one can treat them as a set of vectors, r_1, r_2 and r_3, that are directed
   along the bonding axes of AX_3. These vectors must form a basis for a
   representation of the group (Fig. 4.5).

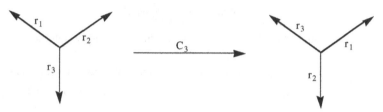

■ **FIGURE 4.5** Vector representation for the D_{3h} point group.

a. Consider a C_3 rotation.

$$C_3 \begin{pmatrix} r_1 \\ r_2 \\ r_3 \end{pmatrix} = \begin{pmatrix} r_3 \\ r_1 \\ r_2 \end{pmatrix} = \begin{pmatrix} 0 & 0 & 1 \\ 1 & 0 & 0 \\ 0 & 1 & 0 \end{pmatrix} \begin{pmatrix} r_1 \\ r_2 \\ r_3 \end{pmatrix} \chi(C_3) = 0$$

b. Consider a C_2 rotation (Fig. 4.6).

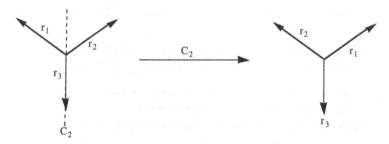

■ **FIGURE 4.6** Vector representation for σ bonding under C_2 axis of rotation.

$$C_2 \begin{pmatrix} r_1 \\ r_2 \\ r_3 \end{pmatrix} = \begin{pmatrix} r_2 \\ r_1 \\ r_3 \end{pmatrix} = \begin{pmatrix} 0 & 1 & 0 \\ 1 & 0 & 0 \\ 0 & 0 & 1 \end{pmatrix} \begin{pmatrix} r_1 \\ r_2 \\ r_3 \end{pmatrix} \chi(C_2) = 1$$

c. If one looks at the characters for the other symmetry operations, a very simple relationship is found. **The character of the representations is the number of vectors that do not shift during the symmetry operation.**

The characters are: $\chi(E) = 3$; $\chi(C_3) = 0$; $\chi(C_2) = 1$, $\chi(\sigma_v) = 1$, $\chi(\sigma) = 3$; $\chi(S_3) = 0$.

2. π orbitals in AX_3

    a. π Orbitals are treated as vectors that are perpendicular to the σ bonding axis.

    b. There are two sets of π orbitals, one perpendicular to the molecular plane and another in the molecular plane, as shown in Fig. 4.7.

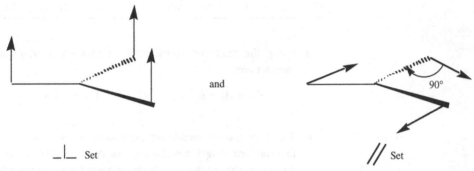

and

$\underline{\perp}$ Set $/\!/$ Set

■ **FIGURE 4.7** Vector representations for π bonding in D_{3h} point group.

There are three possibilities when one carries out the symmetry operations.

    i. A vector could shift; it contributes zero to the character of the representation.

    ii. A vector could remain unchanged; it will contribute $+1$ to the character.

    iii. A vector could change directions; it will contribute -1 to the character.

    c. The characters for the perpendicular set:

$$\chi(E) = 3; \chi(C_3) = 0; \chi(C_2) = -1, \chi(\sigma_v) = 1, \chi(\sigma) = -3; \chi(S_3) = 0$$

    d. The characters for the parallel set:

$$\chi(E) = 3; \chi(C_3) = 0; \chi(C_2) = -1, \chi(\sigma_v) = 1, \chi(\sigma) = 3; \chi(S_3) = 0$$

3. Use the D_{3h} character table (Table 4.5) to construct the molecular orbitals (MOs) for AX_3.

Table 4.5 Character Table for a D_{3h} Point Group

D_{3h}	E	$2C_3$	$3C_2$	σ_h	$2S_3$	$3\sigma_v$		
A_1'	1	1	1	1	1	1		x^2+y^2, z^2
A_2'	1	1	−1	1	1	−1	R_z	
E'	2	−1	0	2	−1	−1	(x, y)	(x^2-y^2, xy)
A_1''	1	1	1	−1	−1	−1		
A_2''	1	1	−1	−1	−1	1	z	
E''	2	−1	0	−2	1	0	(R_x, R_y)	(xz, yz)

a. Using the reduction equation, the irreducible representations in each set are:

$$\Gamma_\sigma = A_1' + E' \quad \Gamma_\perp = A_2'' + E'' \quad \Gamma_{//} = A_2' + E'$$

b. These are the irreducible representations of the sets of X orbitals that can form σ and π bonds with the A in AX_3. These are also the symmetries of the orbitals that A must furnish to bond with the X's.

c. The set of atomic orbitals on the central atom that are symmetry allowed for σ and π bonding would be:

σ bonding A_1': s; a_{z^2} E': (p_x, p_y); $(d_{xy}, d_{x^2-y^2})$
π bonding $\perp A_2''$: p_z E'': (d_{xz}, d_{yz})
        $//A_2''$: none E': (p_x, p_y); $(d_{xy}, d_{x^2-y^2})$

If A is a main group element, it will use s and p orbitals; if it is a transition metal, d orbitals can also be used.

d. Arriving at the sets of atomic orbitals on the X atoms that can be used is more involved. To do this, first define a set of local x, y, z coordinates on each X atom, with the $+x$ direction of each pointing to the central A atom. These local coordinates are defined in the directions of the vectors used to distinguish σ from π bonding (Fig. 4.8).

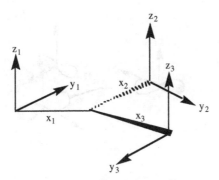

■ FIGURE 4.8 Vector representations to distinguish σ from π bonding.

e. Each X atom will have a totally symmetric s orbital and a set of p_x, p_y, and p_z orbitals. From what has gone before, the s and p_x orbitals are involved in σ bonding while the p_z and p_y orbitals are involved in π bonding (p_z in the $\perp$ set of π bonds and p_y in the $\parallel$ set of π bonds).

Let us consider the linear combinations of X orbitals associated with the one-dimensional irreducible representations.

f. When the projection operator $\hat{P}^j$ is applied to ϕ_1, where ϕ_1 is orbital one in either the σ set (s's and p_x's), the following linear combination is obtained.

$$(\chi_E)\phi_1 + (\chi_{C_3})\phi_3 + (\chi_{C_3})\phi_2 + (\chi_{C_2})\phi_1 + (\chi_{C_2})\phi_3 + (\chi_{C_2})\phi_2 + (\chi_{\sigma_h})\phi_1$$
$$+ (\chi_{S_3})\phi_3 + (\chi_{S_3})\phi_2 + (\chi_{\sigma_v})\phi_1 + (\chi_{\sigma_v})\phi_3 + (\chi_{\sigma_v})\phi_2$$

i. The $\pi_{\parallel}$ set p_y's give:

$$(\chi_E)\phi_1 + (\chi_{C_3})\phi_3 + (\chi_{C_3})\phi_2 - (\chi_{C_2})\phi_1 - (\chi_{C_2})\phi_3 - (\chi_{C_2})\phi_2 + (\chi_{\sigma_h})\phi_1$$
$$+ (\chi_{S_3})\phi_3 + (\chi_{S_3})\phi_2 - (\chi_{\sigma_v})\phi_1 - (\chi_{\sigma_v})\phi_3 - (\chi_{\sigma_v})\phi_2$$

Here note that the C_2 operation carried out on ϕ_1 gives $-\phi_1$.

g. The $\pi_{\perp}$ set (p_z's) gives:

$$(\chi_E)\phi_1 + (\chi_{C_3})\phi_3 + (\chi_{C_3})\phi_2 - (\chi_{C_2})\phi_1 - (\chi_{C_2})\phi_3 - (\chi_{C_2})\phi_2 - (\chi_{\sigma_h})\phi_1$$
$$- (\chi_{S_3})\phi_3 - (\chi_{S_3})\phi_2 - (\chi_{\sigma_v})\phi_1 - (\chi_{\sigma_v})\phi_3 - (\chi_{\sigma_v})\phi_2$$

h. The SALCs for the totally symmetric A_1' irreducible representation are shown in Fig. 4.9. Using the projection operator on orbital ϕ_1,

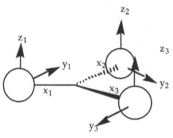

■ **FIGURE 4.9** Orientations of s or d_{z^2} orbitals of the central atom.

$$^{\wedge}P^{A_1'}\phi_1 = \phi_1 + \phi_3 + \phi_2 + \phi_1 + \phi_3 + \phi_2 + \phi_1 + \phi_3 + \phi_2 + \phi_1 + \phi_3 + \phi_2$$
$$= 4(\phi_1 + \phi_2 + \phi_3)$$

The normalized SALC would be $\frac{1}{\sqrt{3}}(\phi_1 + \phi_2 + \phi_3)$. These could interact with the central atom's s or d_{z^2} orbitals.

i. The A_2'' irreducible representation in the $\perp$ π bonding set. Using the projection formula as earlier, the SALC would also be $\frac{1}{\sqrt{3}}(\phi_1 + \phi_2 + \phi_3)$. This set could interact with the central atom's p_z orbital (Fig. 4.10).

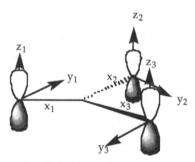

■ **FIGURE 4.10** Orientations of p_z orbitals of the central atom.

j. The A_2' irreducible representation in the // π set (p_y's). Again, using the projection formula as earlier, the SALC would also be $\frac{1}{\sqrt{3}}(\phi_1 + \phi_2 + \phi_3)$. There are no orbitals on the central atom that could interact with this symmetry orbital (Fig. 4.11).

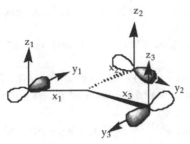

■ FIGURE 4.11 Orientations of p_y orbitals of the central atom.

4. There are two symmetry orbitals corresponding to each E irreducible representation. Rather than use the complete D_{3h} character table to arrive at the SALCs, it is easier to use just the C_3 operations in conjunction with the C_3 character table. It should be noted that the C_3 operations will interchange all the X orbitals. The C_3 character table is shown in Table 4.6.

Table 4.6 Character Table for C_3 Operations

C_3	E	C_3	C_3^2		$\varepsilon = \exp(2\pi i/3)$
A	1	1	1	z, R_z	$x^2 + y^2, z^2$
E	$\begin{cases} 1 \\ 1 \end{cases}$	$\begin{matrix} \varepsilon \\ \varepsilon* \end{matrix}$	$\left.\begin{matrix} \varepsilon* \\ \varepsilon \end{matrix}\right\}$	$(x, y)\ (R_x, R_y)$	$(x^2 - y^2, xy)\ (yz, xz)$

The two-dimensional representation is separated into two one-dimensional ones with imaginary characters. They are used in the same way as other characters. Applying the projection operator to ϕ_1 one obtains:

$$^\wedge P^j \phi_1 = (\chi_E)\phi_1 + \left(\chi_{C_3}\right)\phi_3 + \left(\chi_{C_3^2}\right)\phi_2$$

a. The A representation gives the SALC of $\frac{1}{\sqrt{3}}(\phi_1 + \phi_2 + \phi_3)$, which is the same as obtained by using the D_{3h} point group.

b. The E representation gives the following two linear combinations $(\phi_1 + \varepsilon\,\phi_3 + \varepsilon*\phi_2)$ and $(\phi_1 + \varepsilon*\,\phi_3 + \varepsilon\,\phi_2)$ and the SALCs can be obtained by taking linear combinations of the two. One usually takes the linear combinations so as to eliminate the imaginary parts. Recall the identities from trigonometry

$\varepsilon = \exp(2\pi i/3) = \cos(2\pi/3) + i\sin(2\pi/3)$ and $\varepsilon^* = \cos(2\pi/3) - i\sin(2\pi/3)$.

$\therefore \ \varepsilon + \varepsilon^* = 2\cos(2\pi/3) = -1$ and $(\varepsilon - \varepsilon^*)/i = 2\sin(2\pi/3) = \sqrt{3}$

  i. $(\phi_1 + \varepsilon\,\phi_3 + \varepsilon^*\phi_2) + (\phi_1 + \varepsilon^*\,\phi_3 + \varepsilon\,\phi_2) = 2\phi_1 - \phi_2 - \phi_3$

  ii. $[(\phi_1 + \varepsilon\,\phi_3 + \varepsilon^*\phi_2) - (\phi_1 + \varepsilon^*\,\phi_3 + \varepsilon\,\phi_2)]/i = \sqrt{3}\,\phi_2 - \sqrt{3}\,\phi_3$ or $\phi_2 - \phi_3$

c. the SALCs are $\psi^{E(1)} = \frac{1}{\sqrt{6}}(2\phi_1 - \phi_2 - \phi_3)$, and

$\psi^{E(2)} = \frac{1}{\sqrt{2}}(\phi_2 - \phi_3)$

5. The SALCs for the E irreducible representations are as follows.

a. E' in the σ bonding set of the s orbitals (Fig. 4.12).

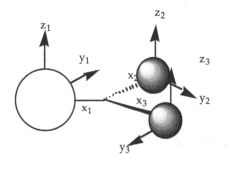 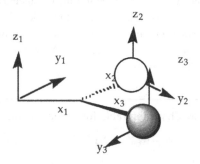

$$\Psi_s^{E'}(1) \qquad\qquad \Psi_s^{E'}(2)$$

FIGURE 4.12 σ bonding set of s orbitals.

These can interact with the p_x and p_y orbitals of a main group central atom as shown in Fig. 4.13.

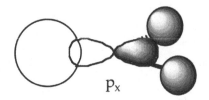 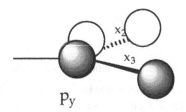

$$P_x \qquad\qquad\qquad P_y$$

FIGURE 4.13 Interactions of s orbital with p_x and p_y orbitals of the central atom.

b. E' in the σ set of the p_x orbitals could also interact with the p_x and p_y orbitals of a main group central atom (Fig. 4.14).

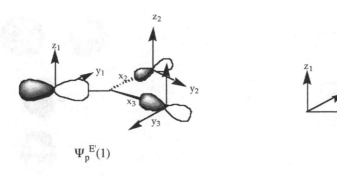

$\Psi_p^{E'}(1)$

$\Psi_p^{E'}(2)$

■ **FIGURE 4.14** Interactions of σ orbital with p_x and p_y orbitals of the central atom.

If the central atom is a main group element, it has no orbitals that could interact with the E'' X SALCs. If it were a transition metal, its d_{xz} and d_{tz} orbitals could interact.

3.3 Molecular orbital correlation diagram for trigonal planar structure (BF_3)

1. BF_3 is a planar triangular molecule belonging to the D_{3h} point group.
   a. Group theory can allow one to choose the B and F atomic orbitals that are symmetry matched to form molecular orbitals. These were derived earlier.
   b. The extent to which they will interact and the composition of the MOs in terms of the SALC orbitals will depend on the relative energy of the B and F orbitals involved.
   c. Recall that only orbitals of comparable energies can form molecular orbitals. This usually means that only valence orbitals are delocalized. That is, even though the F's and the B have 1s orbitals, we may consider them as core electrons that remain on the individual atoms. This is an approximation, and in ab initio calculations they are considered. However, for qualitative bonding descriptions, only valence electrons need be considered.
   d. From symmetry considerations, only one-dimensional irreducible representations $(A'_1, A'_2$ and $A''_2)$ and two-dimensional irreducible representations $(E'$ and $E'')$ are found. Also note that the B has no atomic orbitals with E'' or A'_2 symmetries.
   e. Also recall that the bonding MOs will have more of the character of the low-energy atomic orbitals, whereas the antibonding will be more similar to the higher energy input orbitals.
2. Correlation Diagram for BF_3.
   The following points must be considered before a correlation diagram for BF_3 is drawn (Fig. 4.15).

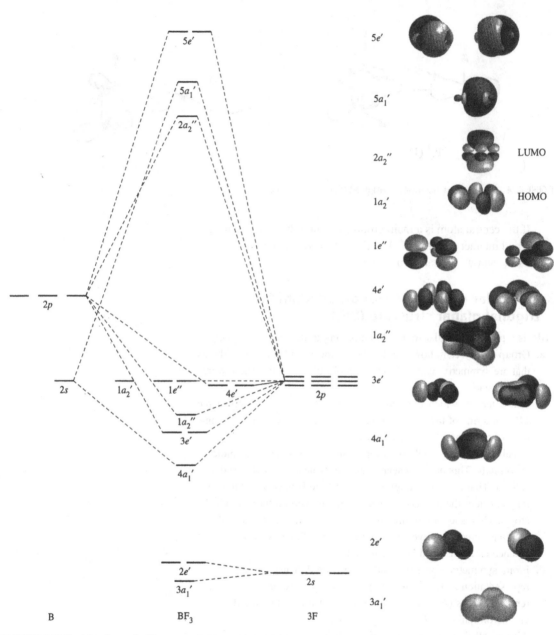

■ **FIGURE 4.15** Correlation diagram for BF₃.

a. BF_3 has 24 valence electrons and 16 valence orbitals. Instead of solving a 16×16 determinate, the use of group theory has simplified it to just several pairwise combinations.

b. Note that the 2s fluorine orbitals are so stable compared with the B orbitals that they interact very little and can be considered as essentially nonbonding; only the p orbitals interact.

c. Also note that from symmetry, the A_2' and E'' F orbitals are nonbonding, as is the e' fluorine SALC arising from their p_y orbitals.

4. EXAMPLES OF OTHER STRUCTURES

4.1 Square planar ML_4

1. Point group is D_{4h}.

2. Set up a coordinate system for the central M atom such that the L groups are along the $+$ and $-x$ axis and the $+$ and $-y$ axis. Use a local coordinate system for each L group in which its x axis is directed toward the central M and the z axis is perpendicular to the ML_4 plane.

3. The irreducible representations contained in the reducible representation for σ and π bonding are:

$$\Gamma_\sigma = A_{1g} + B_{1g} + E_u \quad \Gamma_{\pi\perp} = A_{2u} + B_{2u} + E_g \quad \Gamma_{\pi(//)} = A_{2g} + B_{2g} + E_u$$

4. The M orbitals that could bond:

$$\sigma := s, \, p_x, p_y, d_{z^2}, d_{x^2-y^2} \quad \pi_\perp = p_z, d_{xz}, d_{yx} \quad \pi_{//} = p_x, p_y, d_{xy}$$

5. The SALC orbitals of the L ligands.

a. For the ligands' σ orbitals, $\psi_{A_{1g}} = \frac{1}{2}[\phi_1 + \phi_2 + \phi_3 + \phi_4]$, $\psi_{B_{1g}} = \frac{1}{2}[\phi_1 - \phi_2 + \phi_3 - \phi_4]$, $\psi_{E_u(1)} = \frac{1}{\sqrt{2}}[\phi_1 - \phi_3]$, and $\psi_{E_u(2)} = \frac{1}{\sqrt{2}}[\phi_2 - \phi_4]$, where the ϕ's are the ligand's σ orbitals (s, p_x, or σ lone pair orbitals).

b. For the ligands $\pi_\perp$ orbitals: $\psi_{A_{2u}} = \frac{1}{2}[\phi_1 + \phi_2 + \phi_3 + \phi_4]$, $\psi_{Eg(1)} = \frac{1}{\sqrt{2}}[\phi_1 - \phi_3]$, and $\psi_{E_g(2)} = \frac{1}{\sqrt{2}}[\phi_2 - \phi_4]$, where the ϕ's are the ligand's π orbitals that are perpendicular to the molecular plane (p_z or π molecular orbitals).

c. For the ligands $\pi_\perp$ orbitals: $\psi_{B_{2g}} = \frac{1}{2}[\phi_1 - \phi_2 + \phi_3 - \phi_4]$, $\psi_{Eu(1)} = \frac{1}{\sqrt{2}}[\phi_1 - \phi_3]$, and $\psi_{Eu(2)} = \frac{1}{\sqrt{2}}[\phi_2 - \phi_4]$, where the ϕ's are the ligand's π orbitals that are parallel to the molecular plane (p_y or π molecular orbitals).

6. In a molecule such as XeF_4, the Xe would use its s and p orbitals (vacant d orbitals need not be involved). In $[Ni(CN)_4]^{2-}$ the Ni would use its s, p, and d orbitals.

4.2 **The π orbitals of the cyclopentadienide ion, $[C_5H_5]^-$**

1. The ion is planar and in the D_{5h} point group. Each C has a p_π orbital that is perpendicular to the molecular plane. These will combine to form five π MOs.

2. Using the C_5 group, $\Gamma_\pi = A + E_1 + E_2$. The MOs are:

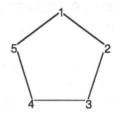

$$\Psi_A = \sqrt{\frac{1}{5}}[\phi_1 + \phi_2 + \phi_3 + \phi_4 + \phi_5]$$

$$\Psi_{E_1(a)} = \sqrt{\frac{2}{5}}[\phi_1 + 0.309\phi_2 - 0.809\phi_3 - 0.809\phi_4 + 0.309\phi_5]$$

$$\Psi_{E_1(b)} = \sqrt{\frac{2}{5}}[0.951\phi_2 + 0.588\phi_3 - 0.588\phi_4 - 0.951\phi_5]$$

$$\Psi_{E_2(a)} = \sqrt{\frac{2}{5}}[0.588\phi_2 - 0.951\phi_3 + 0.951\phi_4 - 0.588\phi_5]$$

$$\Psi_{E_2(b)} = \sqrt{\frac{2}{5}}[\phi_1 - 0.809\phi_2 + 0.309\phi_3 + 0.309\phi_4 - 0.809\phi_5]$$

The MOs as viewed from the perspective of a bonding metal (from above the molecular plane) are shown in Fig. 4.16.

Ψ_A
$E = \alpha+2\beta$

$\Psi_{E_1(a)}$ $\Psi_{E_1(b)}$
$E = \alpha+0.618\beta$

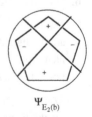
$\Psi_{E_2(a)}$ $\Psi_{E_2(b)}$
$E = \alpha-1.618\beta$

■ **FIGURE 4.16** Molecular orbitals of cyclopentadienide ion, $[C_5H_5]^-$.

5. MOLECULAR SPECTROSCOPY

5.1 Types of molecular motion

1. Consider a molecule composed of N atoms.
   a. Before bonding together, each atom can move in three dimensions in space, each has three degrees of translational freedom, giving a total of 3N degrees of freedom.
   b. After the atoms are bonded together in a molecular unit, there are still 3N degrees of freedom that are expressed in three types of molecular motion.
2. Types of molecular motion.
   a. Translational motion of the molecule. The center of mass of the molecule can move in the x, y, z directions in space. The motion is not quantized and its energy is its kinetic energy. According to the Kinetic Molecular Theory, this energy is equal to $\frac{3}{2}RT$ ($\frac{1}{2}RT$ for each degree of translational freedom).
   b. Rotational motion. The molecule can rotate about its axes.
      i. Nonlinear molecules can rotate about their x, y, and z axes, giving three degrees of rotational freedom.
      ii. Linear molecules can rotate about two axes, giving two degrees of rotational freedom.
      iii. Rotational motion is quantized, and the transitions from one rotational energy state to another can be observed using microwave spectroscopy. Analysis of the microwave spectrum can lead to experimental values of the equilibrium internuclear distance.
   c. Vibrational motion. A molecule's bond can be stretched and compressed as can its bond angles.
      i. All the other degrees of molecular motion can be resolved into vibrational motion.
      ii. For a nonlinear molecule containing N atoms, there are 3N−6 degrees of vibrational freedom, whereas for a linear molecule, there are 3N−5 degrees.

5.2 Normal modes of vibration and symmetry

1. The vibrational contortions of a molecule can be broken down into a set of 3N−6 (or 3N−5) specific oscillatory motions called **normal modes**. These consist of sets of symmetric and asymmetric bond stretching and angle bending motions. In general, bond stretching vibrations are higher in energy than the angle deformations.
2. One can use group theory to describe the normal modes of vibration.

3. Consider the BF_3 molecule. It has $3(4) - 6 = 6$ normal modes of vibration. The reducible representation for the molecule's 3N degrees of freedom of molecular motion can be obtained by assigning a local coordinate system and x, y, z directional vectors to each atom as shown in Fig. 4.17. By convention, the z direction is taken parallel to the major symmetry, in this case the C_3 axis.

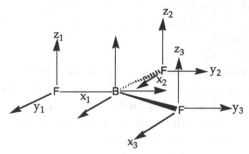

■ **FIGURE 4.17** Vectors at the C_3 axis of each atom in BF_3 molecule.

a. As was the case in constructing the π and σ MOs, each vector that remains unchanged in a symmetry operation contributes 1 to the character of the irreducible representation, each vector that changes direction contributes -1, whereas a vector that is shifted contributes zero.

b. Since the local coordinates might not coincide with the symmetry elements of the molecule, there is the possibility of a symmetry operation transforming the coordinates into a linear combination of its original x, y, and z. This is depicted in Fig. 4.18 for a C_3 rotation of the local axes on the B atom.

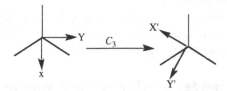

■ **FIGURE 4.18** C_3 axis of rotation at the central boron atom.

We saw earlier that the matrix that would transform the coordinates is

$$C_3 = \begin{pmatrix} -\dfrac{1}{2} & \dfrac{\sqrt{3}}{2} & 0 \\ -\dfrac{\sqrt{3}}{2} & -\dfrac{1}{2} & 0 \\ 0 & 0 & 1 \end{pmatrix}. \text{ Note that the character of this matrix,}$$

$\chi(C_3)$ is zero.

c. By considering the characters of the matrices that will carry out the transformation of the unshifted atoms on the various symmetry operations, Table 4.7 can be obtained. This table is independent of the molecule's point group.

Table 4.7 Contribution to χ of Each Unshifted Atom

R	Contribution	R	Contribution	R	Contribution
E	3	C_6	2	S_4	-1
C_2	-1	Σ	1	S_6	0
C_3	0	I	-3	S_5	-0.382
C_4	1	S_3	-2	S_n^k	$-1 + 2\cos(2\pi k/n)$
C_5	1.618	$C_5^2 = C_5^3$	-0.618	C_n^k	$1 + 2\cos(2\pi k/n)$

d. From Table 4.7, the characters for the irreducible representations, shown in Table 4.8, can be obtained.

Table 4.8 Characters for Irreducible Representations of BF_3 Molecule

R	E	$2C_3$	$3C_2$	σ_h	$2S_3$	$3\sigma_v$
No. of unshifted atoms	4	1	2	4	1	2
Contribution per atom	3	0	-1	1	-2	1
Γ_{red}	12	0	-2	4	-2	2

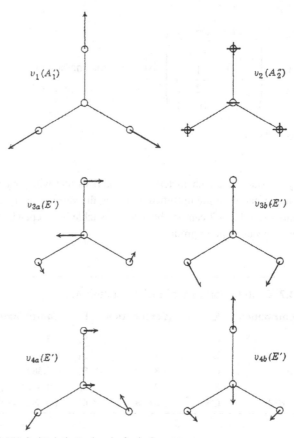

■ **FIGURE 4.19** Predicted vibrational modes for the D_{3h} point group.

    i. Using the reduction formula and the D_{3h} character table, the irreducible representation for molecular motion can be shown to be:

$$\Gamma_{red} = A_1' + A_2' + 2A_2'' + 3E' + E''.$$

    ii. Subtracting out those irreducible representations associated with translation, $(A_2'' + E')$, and rotation, $(A_2' + E'')$, one is left with $\Gamma_{vib} = A_1' + A_2'' + 2E'$.

   iii. For a vibration to be infrared (IR) active, the vibration must lead to a change in the dipole moment of the molecule. Those vibrations that form the basis for the same irreducible representations as the involved coordinates x, y, and z have this property. From the D_{3h} character table these are the A_2' and the E modes.

iv. For a vibration to be Raman active, the vibration must lead to a change in the polarizability of the molecule; such vibrations form the same bases for the same irreducible representations as do the squares and binary products of the coordinates (x^2, y^2, z^2, xy, xz, yz or their combinations). From the D_{3h} character table these would be the A_1' and the E' vibrations.

v. In summary, the BF_3 molecule (as well as the $CO_3{}^{2-}$ and the $NO_3{}^-$ ions) will have three IR absorptions (A_2' and the two doubly degenerate E' modes) and three Raman active vibrations (the A_1' and the two doubly degenerate E' vibrations). Note that the E' modes are both IR and Raman active.

vi. The six normal modes are shown in Fig. 4.19. Each forms the basis for a particular representation of the group, but, with the exception of the symmetric stretch, the relationship between the normal modes is not especially obvious.

e. Note that the E' modes are both IR and Raman active. A review of the different point group character tables will show that if the molecule (or ion) has a center of inversion (i), then the IR and Raman active bands are mutually exclusive. Fig. 4.19 shows the normal mode of vibrations for a planar triangular substance in the D_{3h} point group.

BIBLIOGRAPHY

1. Goodney, D. E. Quantum History Chart. Willamette University *J. Chem. Educ.* **1991,** *68* (6), 473.
2. Garrett, A. B. Quantum Theory: Max Planck. *J. Chem. Educ.* **1963,** *40* (5), 262−263.
3. Dushman, S. Quantum Theory. *J. Chem. Educ.* **1931,** *8* (6), 1075−1112.
4. Hazlehu, T. H. Quantum Numbers and the Periodic Table. *J. Chem. Educ.* **1941,** 580−581.
5. Hall, M. B. Valence Shell Electron Pair Repulsions and the Pauli Exclusion Principle. *J. Am. Chem. Soc.* **1978,** *100* (20), 6333−6338.
6. Michalak, A.; DeKock, R. L.; Ziegler, T. Bond Multiplicity in Transition Metal Complexes: Applications of Two-electron Valence Indices. *J. Phys. Chem.* **2008,** *112,* 7256−7263.
7. Shenkuan, N. The Physical Basis of Hund's Rule. *J. Chem. Educ.* **1992,** *69* (10), 800−803.
8. Campbell, M. L. The Correct Interpretation of Hund's Rule as Applied to "Uncoupled States" Orbital Diagrams. *J. Chem. Educ.* **1991,** *68* (2), 134−135.
9. Walls, J. D.; Heller, E. J. Spin-orbit Coupling Induced Interference in Quantum Corrals. *Nano Lett.* **2007,** *7* (11), 3377−3382.
10. Remenyi, C.; Reviakine, R.; Arbuznikov, A. V.; Vaara, J.; Kaupp, M. Spin-orbit Effects on, Hyperfine Coupling Tensors in Transition Metal Complexes Using Hybrid Density Functionals and Accurate Spin-orbit Operators. *J. Phys. Chem.* **2004,** *108,* 5026−5033.

11. Gillespie, R. J. The Valence-shell Electron-pair Repulsion (VSEPR) Theory of Directed Valency. *J. Chem. Educ.* **1963,** *40* (6), 295–301.
12. Gillespie, R. J. The Electron-pair Repulsion Model for Molecular Geometry. *J. Chem. Educ.* **1970,** *47* (1), 18–23.
13. Hargittai, I. Trigonal-bipyrimidal Molecular Structures and the VSEPR Model. *J. Chem. Educ.* **1982,** *21,* 4335–4337.
14. Smith, D. Valence, Covalence, Hypervalence, Oxidation State, and Coordination Number. *J. Chem. Educ.* **2005,** *82* (8), 1202–1204.
15. Brown, M. F.; Cook, B. R.; Sloan, T. E. Stereochemical Notation in Coordination Chemistry: Mononuclear Complexes of Coordination Numbers Seven, Eight, and Nine. *J. Chem. Educ.* **1978,** *17* (6), 1563–1568.
16. Meyers, R. T. Rules for Coordination Number of Metal Ions. *J. Chem. Educ.* **1981,** *58* (9), 681.
17. Carvajal, M. A.; Novoa, J. J.; Alvarez, S. Choice of Coordination Number in d^{10} Complexes of Group 11 Metals. *J. Am. Chem. Soc.* **2004,** *126,* 1465–1477.
18. Muetteries, E. L.; Wright, C. L. Chelate Chemistry. III. Chelates of High Coordination Number. *J. Am. Chem. Soc.* **1965,** *87* (21), 4706–4717.
19. Stewart, A. W. Stereochemistry. *J. Am. Chem. Soc.* **1908,** *30* (8), 1321–1322.
20. Mercandelli, P.; Sironi, A. Ligand Stereochemistry of Metal Clusters Containing pi-Bonded Ligands. *J. Am. Chem. Soc.* **1996,** *118,* 11548–11554.
21. Bent, R. L. Aspects of Isomerism and Mesomerism II. Structural Isomerism. *J. Chem. Educ.* **1953,** *30* (6), 284–290.
22. Zeldin, M. An Introduction to Molecular Symmetry and Symmetry Point Groups. *J. Chem. Educ.* **1966,** *43* (11), 17–20.
23. Chen, G. Symmetry Elements and Molecular Achirality. *J. Chem. Educ.* **1992,** *69* (2), 159.
24. Senior, J. K. Some Aspects of Molecular Symmetry. *J. Org. Chem.* **1936,** *1* (3), 254–264.
25. Carlos, J. L. Molecular Symmetry and Optical Inactivity. *J. Chem. Ed.* **1968,** *45* (4), 248–251.
26. Brown, R. J. C.; Brown, R. F. C. Melting Point and Molecular Symmetry. *J. Chem. Educ.* **2000,** *77* (6), 724–731.
27. Li, W. Identification of Molecular Point Groups. *J. Chem. Educ.* **1993,** *70* (6), 485–487.
28. Baraldi, I.; Vanossi, D. On the Character Tables of Finite Point Groups. *J. Chem. Educ.* **1997,** *74* (7), 806–809.
29. Orchin, M. M.; Jaffe, H. H. Symmetry, Point Groups, and Character Tables. *J. Chem. Educ.* **1970,** *47* (5), 372–377.
30. Purcell, K. F.; Kotz, J. C. *An Introduction to Inorganic Chemistry;* Saunders College Publishing: Philadelphia, 1980, 637 pp.
31. Ivanov, J. Molecular Symmetry Perception. *J. Chem. Inf. Comput. Sci.* **2004,** *44* (2), 596–600.
32. Cohen, I.; Del Bene, J. Hybrid Orbitals in Molecular Orbital Theory. *J. Chem. Educ.* **1969,** *46* (8), 487–492.
33. Harrison, J. F.; Lawson, D. B. Some Observations on Molecular Orbital Theory. *J. Chem. Educ.* **2005,** *82* (8), 1205–1209.

34. Gray, H. B. Molecular Orbital Theory for Transition Metal Complexes. *J. Chem. Educ.* **1964,** *41* (1), 1−12.

35. Gimarc, B. M. Applications of Qualitative Molecular Orbital Theory. *Acc. Chem. Res.* **1974,** *7.*

36. Pearson, R. G. Physical and Inorganic Chemistry. *J. Am. Chem. Soc.* **1969,** *91* (8), 4947−4955.

37. I'Haya, Y. Some Studies in Molecular Orbital Theory. *J. Chem. Educ.* **1958,** *81,* 6120−6127.

38. Sannigrahi, A. B.; Kar, T. Molecular Orbital Theory of Bond Order and Valency. *J. Chem. Educ.* **1988,** *65* (8), 674−676.

39. Couty, M.; Hall, M. B. Generalized Molecular Orbital Theory II. *J. Phys. Chem.* **1997,** *101* (37), 6936−6944.

40. Ewing, G. E. Spectroscopic Studies of Molecular Motion in Liquids. *Acc. Chem. Res.* **1969,** *2* (6), 168−174.

41. Osguthorpe, P.; Osguthorpe, D. Extraction of the Energetics of Selected Types of Motion from Molecular Dynamics Trajectories by Filtering. *Biochemistry* **1990,** *29* (36), 8223−8228.

Part **2**

Advanced Topics-1: Introduction to Ligands and Metal Complexes

Chapter 5

Ligands and *d*-Block Metal Complexes

1. INTRODUCTION: IS IT NECESSARY TO KNOW ABOUT LIGANDS AND METAL COMPLEXES?

It is a fact that the inhalation of gaseous carbon monoxide (CO) or the ingestion of sodium cyanide (NaCN), either as a solid or in solution, can lead to death. The question arises as to what is happening in terms of chemical interactions. In the language of biochemistry, blood contains iron atoms, which are linked to a binding protein known as *hemoglobin/myoglobin*, with retention of a couple of unoccupied binding sites. When we breathe in, one oxygen molecule temporarily binds to an iron atom in the hemoglobin/myoglobin, with proteins occupying four of the six binding sites of the central iron atom, before moving on to other blood cells to eventually reach the heart to facilitate blood purification. In other words, our continuous breathing of oxygen is necessary for sustaining a normal healthy life. However, the presence of carbon monoxide or cyanide ion in the blood leads to these species occupying the required binding sites on the central iron atom, such that the sites are unavailable for oxygen to bind. Therefore, breathing is forced to stop, leading to death. Although this may seem to be part of a biochemical process, the same phenomenon can be explained in the language of coordination chemistry where one oxygen molecule is bound by a coordinate covalent bond to an iron atom (Fe(II)) in the *heme group*. Heme is a square planar molecule containing four *pyrrole* groups called ligands, whose nitrogens form coordinate covalent bonds with four of the iron's six available positions and a remaining orbital is used for binding oxygen. It is empty in the nonoxygenated forms of hemoglobin and myoglobin. Oxygen binds weakly (called a weak-field ligand), whereas the neutral $C\equiv O$ molecule and the negatively charged cyanide ion $(C\equiv N)^-$ are members of a strong-field ligand family. They bind very strongly to the six-coordinated central metal atom. Consequently, with the prior occupation of the binding sites, oxygen cannot enter the heme group and breathing stops.

This leads to terminologies such as "complex ion", "ligands", "coordination number", etc. Almost all of the transition (*d*-block) metals have the ability to

Advanced Inorganic Chemistry. http://dx.doi.org/10.1016/B978-0-12-801982-5.00005-9

form *complex ions* with the metal ion at its center and with a number of other molecules or ions, known as *ligands*, surrounding it with coordinate or dative covalent bonds formed by the ligands donating their lone pairs of electrons. In a simple language of chemistry, all ligands function as *Lewis bases*. The *coordination number* is the number of dative covalent bonds to the central metal ion.

Other common complex ions are present in *vitamin B12*, *chlorophyll*, and *some dyes* and *pigments*. One major use of complexes or coordination compounds is in *homogeneous catalysis* for the production of organic compounds. Thus, they are useful compounds both in nature and industry. Nevertheless, it is essential for us to make the connection between coordination chemistry and processes that occur in nature and in our lives.

2. TRANSITION METALS
2.1 Electronic structures and oxidation states

1. In forming the cations of the elements shown in Table 5.1, 4s electrons are removed first before losing 3d electrons.

Table 5.1 Electronic Structure of Elements in Groups 3 Through 12

Group	First Row Configuration	Valence Configuration
3	$_{21}Sc[Ar]^{18}3d^14s^2$	d^1s^2
4	$_{22}Ti[Ar]^{18}3d^24s^2$	d^2s^2
5	$_{23}V[Ar]^{18}3d^34s^2$	d^3s^2
6	$_{24}Cr[Ar]^{18}3d^54s^1$	d^5s^1
7	$_{25}Mn[Ar]^{18}3d^54s^2$	d^5s^2
8	$_{26}Fe[Ar]^{18}3d^64s^2$	d^6s^2
9	$_{27}Co[Ar]^{18}3d^74s^2$	d^7s^2
10	$_{28}Ni[Ar]^{18}3d^84s^2$	d^8s^2
11	$_{29}Cu[Ar]^{18}3d^{10}4s^1$	$d^{10}s^1$
12	$_{30}Zn[Ar]^{18}3d^{10}4s^2$	$d^{10}s^2$

$_{27}Co^{2+}[Ar]^{18}3d^7$, $Co^{3+}[Ar]^{18}3d^6$, $_{28}Ni^{2+}[Ar]^{18}3d^8$, $_{24}Cr^{2+}[Ar]^{18}3d^4$

As more d electrons become involved in bonding, variable valences are found (exceptions are the group 3 and 12 metals).

2. Consider some compounds of iron where the oxidation state of the metal varies from -2 to $+6$ depending on the complex (Table 5.2).

Table 5.2 Compounds of Iron and the Corresponding Oxidation States

Compound	Oxidation State	Compound	Oxidation State
$Fe(CO)_4^{2-}$	$-II$	Common	$+II$
$Fe(bipy)_3^-$	$-I$	Common	$+III$
$Fe(CO)_5$	0	FeO_4^{4-}	$+IV$
$[Fe(H_2O)_5NO]^{2+}(NO^+)$	$+I$	FeO_4^{3-}, FeO_4^{2-}	$+V, +VI$

3. The oxidation states of the transition metals are functions of the complex in which the metal is involved. These compounds are called *coordination complexes*; the chemistry of the transition metals is the chemistry of coordination compounds.

2.2 **Coordination compounds**

Alfred Werner performed systematic studies to describe bonding in coordination compounds, especially when organic bonding theory and simple ideas of ionic charges were not sufficient. Thus, there have been two types of bonding: (1) primary—positive charge of the metal ion is balanced by negative ions in the compound and (2) secondary—molecules or ion (ligands) are attached directly to the metal ion.

1. Lewis acid—base adducts in which the transition metal atom or ion acts as a Lewis acid. The Lewis bases are called *ligands*. Most ligands are anions or neutral molecules. The only common cationic ligand is the nitrosyl group (NO^+). *Coordination number* is the number of σ bonds formed by the metal with the ligands.
2. In addition to σ bonds, the metal and ligands can also participate in π bonding. There are two types of metal—ligand π bonding.
   a. The π bonds formed by the overlap of filled ligand orbitals with vacant metal orbitals.
      i. Ligands such as OH^-, O^{2-}, and F^- can participate in forward (regular) π bonding.
      ii. These should stabilize high oxidation state compounds. Note that the high oxidation states of Fe are found in the iron oxides.

b. The π bonds formed by the overlap of vacant ligand orbitals with filled metal orbitals.

This type of π bonding is called *back π bonding*. Electron density transfer is in the opposite direction from that in σ bonding.

i. This is important in stabilizing low oxidation state complexes.

ii. Examples of ligands that can undergo back π bonding are: CO, NO^+, PR_3, olefins, and polyolefins. Note in the table of Fe complexes the low oxidation state complexes included those with CO and NO^+.

2.3 Ligands (Lewis bases)

1. Monodentate: This is the most common type of ligand.

The Lewis base can form only one σ bond with a particular metal. In most cases, the base site is a lone pair of electrons in a hybridized orbital as in $|NH_3$, H_2O, $|C\equiv O|$, Cl^-, OH^-, $|PR_3$, I^-, F^-, and Br^- ligands.

Olefins and polyolefins can use filled π molecular orbitals to form the primary σ bonds. Consider Zeise's salt, $[PtCl_3C_2H_4]^-$, whose structure was determined to show that the ethene molecule is perpendicular to the plane of the complex (Fig. 5.1).

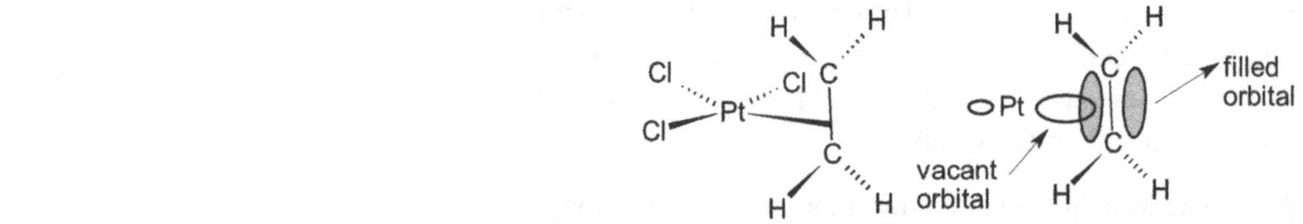

■ **FIGURE 5.1** Molecular structure of $[PtCl_3C_2H_4]^-$ complex ion.

The coordinate covalent bond formed between the ethene and the metal is thought to arise from the overlap of a vacant metal orbital with the filled π olefin orbital.

2. Polydentate ligands.

Polyatomic molecules, with several base sites, can form more than one sigma (σ) bond with a particular metal. The majority of the polydentate ligands have their base sites separated by two atoms so that they form a five-membered ring when they coordinate to the metal. Some molecules have the ability to form four-membered rings, such as oxyanions, but six and higher membered rings do not form easily. Examples are shown in Fig. 5.2.

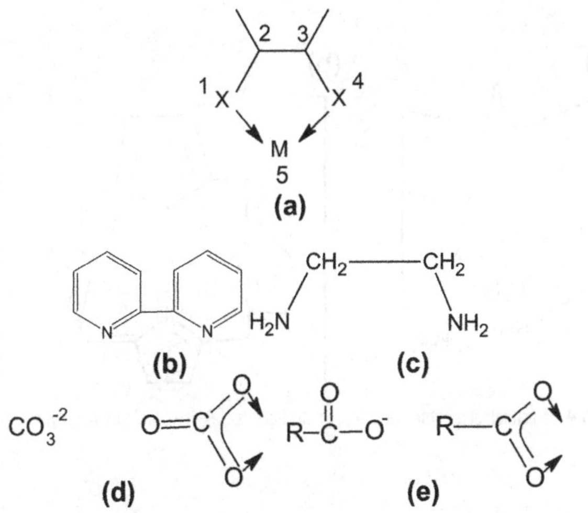

■ **FIGURE 5.2** Common polydentate ligands: (a) general form (X = base site), (b) 2,2′-bipyridine (bipy), (c) ethylene diamine (en), (d) carbonate ion, and (e) alkyl acetate ion.

3. The classification of polydentate ligands.
   a. *Bidentate ligands* form two σ bonds. Examples are (1) ethylene diamine, (2) 2,2′-bipyridine, and (3) carbonate anion (CO_3^{2-}).
   b. *Tridentate ligands* form three σ bonds. An example is $H_2NCH_2CH_2NHCH_2CH_2NH_2$ diethylene triamine (dien). This is a flexible ligand that can link to the metal to give several geometries. Fig. 5.3 shows two possible orientations of the ligand in an octahedral complex.

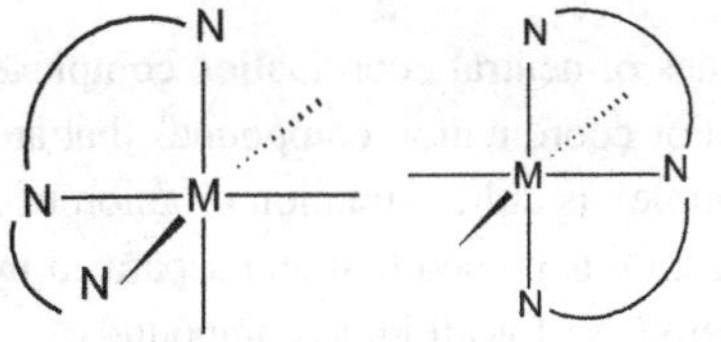

■ **FIGURE 5.3** Two possible orientations of a tridentate ligand.

   c. *Tetradentate ligands* can form four σ bonds. There are several general types.
   Linear, flexible ligand of the type is triethylene tetramine $H_2NCH_2CH_2NHCH_2CH_2NHCH_2CH_2NH_2$ (trien).

Tripodal ligand of the type can be Tren and planar *macrocyclic* ligand of the type is porphyrin, whose structural geometries can be seen in Fig. 5.4.

(a)   **(b)**

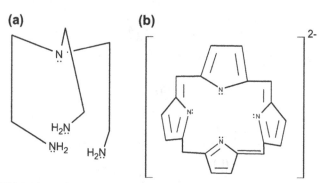

■ **FIGURE 5.4** Structural geometries of tetradentate ligands: (a) Tren and (b) porphyrin.

The planar macrocyclic metalloporphyrin complexes are biologically important molecules.

Examples include hemoglobin, cytochromes (Fe-porphyrins), and chlorophyll (Mg-porphyrin).

d. *Penta- and hexadentate ligands* also form a series of five-membered rings and are macrocyclic in nature.

3. NOMENCLATURE OF COORDINATION COMPOUNDS

The following IUPAC (International Union of Pure and Applied Chemists) rules are used to name the complexes systematically.

Rule 1: The names of neutral coordination complexes are given without spaces. For coordination compounds that are ionic (i.e., the coordination complex is either an anion or anion of an ionic substance), the cation is named first and separated by a space from the anion, as is system case for all ionic compounds.

Rule 2: The name of the coordination compound (neutral, cationic, or anionic) begins with the names of the ligands. The metal is listed next, followed in parentheses by the oxidation state of the metal.

Rule 3: When more than one of a given ligand species is bound to the same metal atom or ion, the number of such bonded ligands is designated by prefixes such as mono, di, tri, tetra, penta, etc.

Rule 4: Neutral ligands, such as pyridine, 2,2′-bipyridine, terpyridine, ethylenediamine, urea, and other neutral species, are given the same name as the uncoordinated molecule, but with spaces omitted.

Rule 5: Anionic ligands are given names that end in the letter "o". For example, when the name of the free, uncoordinated anion ends in "ate", the ligand name is changed to end in "ato".

Rule 6: The ligands are named alphabetically, ignoring the prefixes *bis*, *tris*, etc.

Rule 7: When the coordination entity is either neutral or cationic, the usual name of the metal is used, followed in parentheses by the oxidation state of the metal. However, when the coordination entity is an anion, the name of the metal is altered to end in "ate". This is done for some metals by simply changing the ending "ium" to "ate".

Rule 8: Bridging ligands are designated with the prefix μ (mu). When there are two bridging ligands of the same kind, the prefix μ,μ′ (mu, mu prime) is used. Bridging ligands are listed in order with other ligands, according to Rule 6, and set off with hyphens (-). An important exception arises when the molecule is symmetrical, and a more compact name can then be given by listing the bridging ligand first.

Rule 9: Optical isomers are designated experimentally by the symbols D (dextrorotatory with + sign) or L (levorotatory with a − sign) and named by using R or S, or Λ (lambda for left handed) and Δ (delta for right handed). Geometrical isomers are designated by *cis-* or *trans-* and *mer-* or *fac-*, the latter two standing for meridional or facial, respectively.

Rule 10: Ligands that are capable of linkage isomerism are given specific names for each mode of attachment.

According to these rules, for any salt, name the cation first then the anion independent of which is complex.

Naming complex cations or neutral complexes.

1. Name the ligands first and indicate the number of times each ligand occurs.
    a. Anionic ligands.
        i. If the anion ends in *ide* drop the *ide* and add *o*.
        Examples : $Cl^- = $ chloro $Br^- = $ bromo $OH^- = $ hydroxo
                     $I^- = $ iodo $CN^- = $ cyano $O^{2-} = $ oxo
        Exception : $NH_2^- = $ amido

      ii. If the anion ends in *ate* or *ite*, drop the *e* and add *o*.

Examples :

SO_4^{2-} = sulfato	$C_2O_4^{2-}$ = oxolato
SO_3^{2-} = sulfito	CO_3^{2-} = carbonato
NO_2^- = nitrito (if O bonded)	
SCN^- = thiocyanato (if N bonded)	
NCS^- = isothiocyanato (if S bonded)	
(see Rule#10)	

Exception : NO_2^- = nitrito (if N bonded)

  b. Neutral and cationic ligands.

      i. Use the name of the molecule without alteration.

          Exceptions : H_2O = aquo   CO = carbonyl

                       NH_3 = ammine NO^+ = nitrosyl cation

  c. Use prefixes to indicate the number of times a ligand occurs.

      i. di = 2   tri = 3   tetra = 4   penta = 5   hexa = 6, etc.

      ii. For complex ligands that already have di-, tri-, etc. in their name, use the Greek prefixes and enclose the ligand name in parenthesis.

         bis = 3   tris = 3   tetrakis = 4   pentakis = 5, etc.

  d. Name the ligands in alphabetical order (Rule 6).

2. After naming the ligands and indicating their numbers, give the name of the metal and write its oxidation state (charge) in Roman numerals and enclose in parenthesis.

Examples.

$[Co(NH_3)_4Cl_2]^+$	tetraamminedichlorocobalt(III)
$Fe(CO)_5$	pentacarbonyliron(0)
$[Cr(en)_2I_2]^+$	bis(ethylenediamine)diiodochromium(III)
$[Pt(H_2O)_4](NO_3)_2$	tetraaquoplatinum(II) nitrate

Anionic complexes (Rule 7).

    i. Name and number the ligands in the same way as for cationic complexes.

    ii. Drop the metallic ending of the metal (ium) and add *ate*.

Examples : Scandium   =   scandate
Titanium   =   titanate
Chromium   =   chromate
Zirconium   =   zirconate
Niobium   =   niobate
Ruthenium   =   ruthenate
Palladium   =   palladate
Rhenium   =   rhenate

For other metals, the name is given the ending "ate" and there are several ways in which the name arises. For example,

Manganese   =   manganate
Cobalt   =   cobaltate
Nickel   =   nickelate
Tantalum   =   tantalate
Tungsten   =   tungstate
Platinum   =   platinate

Finally, the names for some metals are based on the Latin name of the element:

Iron   =   ferrate
Copper   =   cuprate
Silver   =   argentite
Mercury(Hydrargyrum)   =   mercurate(hydrargyrate)

iii. Examples:

$Fe(C_2O_4)_3^{3-}$ trioxolatoferrate(III)

$[PtI_4]^{2-}$ tetraiodoplatinate(II)

$[Co(CN)_5OH]^{3-}$ pentacyanohydroxocobaltate(III)

For bridging complexes, Rule 8 will be applied.

i. Use μ (mu) to indicate a bridging group. μ should be repeated for each different bridging ligand. Examples are shown in Fig. 5.5.

(a)   **(b)**

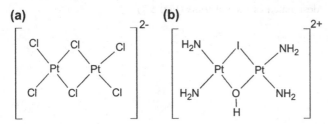

■ **FIGURE 5.5** Structural geometries of (a) dichloroplatinum(II)-μ-dichlorodichloroplatinate(II) and (b) diammineplatinum(II)-μ-iodo-μ-hydroxo-diammineplatinum(II).

4. ISOMERISM IN COORDINATION COMPOUNDS

4.1 Coordination number = 4

1. Tetrahedral complexes.

    a. The only isomerism possible is optical isomerism in complexes having four different ligands surrounding a metal.

    b. These are usually very difficult to resolve into separate isomers since most T_d complexes undergo a very rapid ligand exchange (they are kinetically labile complexes).

2. Square planar complexes.

    a. Consider $[PtCl_4]^{2-}$. This complex is square planar with D_{4h} symmetry; all four Cl^- ligands are equivalent. No isomers are possible (Fig. 5.6a).

    b. Consider $[PtCl_3Br]^{2-}$ anionic complex. Think of it being formed by replacing one Cl^- ligand in $[PtCl_4]^{2-}$ with a Br^-. Since all Cl^-'s are equivalent, it does not matter which Cl^- is replaced. Therefore only one structure exists without any isomeric forms. However, there are two different types of Cl^- ions; the one opposite, or *trans* to the Br^- ion and the two next to, or *cis* to the Br^- ion (Fig. 5.6b).

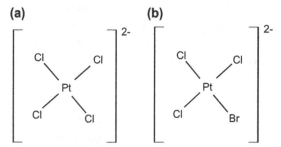

■ **FIGURE 5.6** Structures of (a) tetrachloroplatinate(II) and (b) bromotrichloroplatinate(II) (C_{2v}).

    c. Consider $[PtCl_2Br_2]^{2-}$. Think of this complex as being derived from $[PtCl_3Br]^-$ by replacing a Cl with a Br. Since there are two different types of Cl^-'s, there are two isomers of $[PtCl_2Br_2]^-$, designated as *cis* and *trans* (Fig. 5.7).

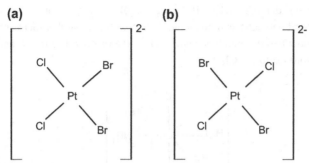

■ **FIGURE 5.7** Structures of (a) *cis*-dibromodichloroplatinate(II) (C_{2v}) and (b) *trans*-dibromodichloroplatinate(II) (D_{2h}).

In the *cis* isomer the two like ligands are next to one another, whereas in the *trans* isomer they are opposite.

d. Consider a square planar complex with four different ligands, [MABCD]. There are three different isomers with A opposite to B, C, or D (Fig. 5.8). Note that the molecular plane is a mirror plane. Therefore none of these isomers are optically active.

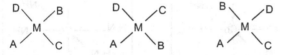

■ **FIGURE 5.8** Three possible geometries of a square planar complex [MABCD].

3. Octahedral complexes.

a. Consider $[Co(NH_3)_6]^{3+}$ (Fig. 5.9) in which all six NH_3's are equivalent. The molecule has octahedral symmetry with one structure without any isomeric forms.

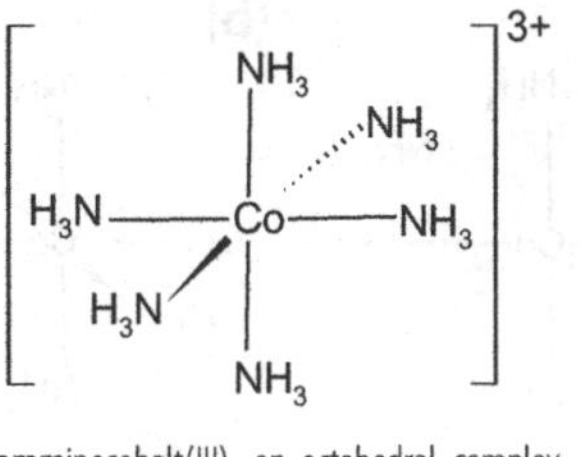

■ **FIGURE 5.9** Structure of hexamminecobalt(III), an octahedral complex.

b. Consider $[Co(NH_3)_5Cl]^{2+}$ (Fig. 5.10); it exhibits only one structural geometry and no isomers are possible. Note that there are now two differently situated NH_3's: the one *trans* to the Cl and the four *cis* to the Cl.

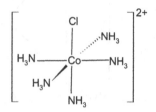

■ **FIGURE 5.10** Structure of pentamminechlorocobalt(III), an octahedral complex.

c. Consider $[Co(NH_3)_4Cl_2]^+$ complex (Fig. 5.11) that exhibits two isomers, *cis* and *trans*, with respect to the positions of the two Cl atoms.

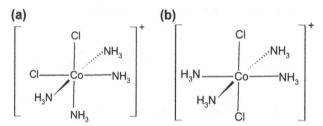

■ **FIGURE 5.11** Structure of *cis*-tetramminedichlorocobalt(III) and *trans*-tetramminedichlorocobalt(III).

d. Consider $[Co(NH_3)_3Cl_3]$ complex (Fig. 5.12) that forms two isomers, the facial (*fac*) (a) and the meridional (*mer*) isomer (b).
e. More complex isomers can exist for octahedral complexes.

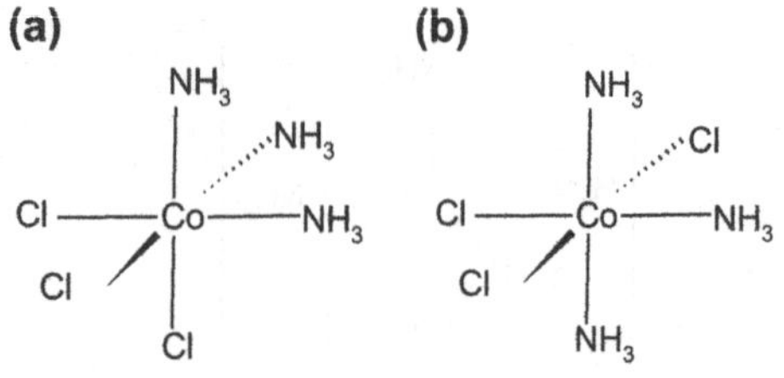

(a)   **(b)**

■ **FIGURE 5.12** Structure of *fac*-triamminetrichlorocobalt(III) and *mer*-triamminetrichlorocobalt(III).

i.   Consider $[Co(NH_3)_2Cl_2Br_2]^-$ anionic complex (Fig. 5.13) in which there are a number of different isomers.

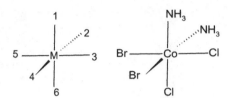

■ **FIGURE 5.13** Structure of 1,2-diammine-4,5-dibromo-3,6-dichlorocobalt(III) anion.

ii. It is essential to number the positions to locate the ligands.

Review of Bonding Theories
for *d*-Block Metal Complexes

1. INTRODUCTION: WHY BONDING THEORIES OF METAL COMPLEXES ARE IMPORTANT?

The most common curiosity among the students is why most transition metal complexes are brightly colored but some are not. Similarly, why do the colors change as the ligands change or when the oxidation state of the metal changes, even for complexes of the same ligand? Is there a relationship between the magnetic property of a complex and the number of unpaired electrons? To find the answer to the last question, we need to know how the number of unpaired electrons can be determined and how it is related to the magnetic moment of a complex. A *Gouy balance* can be used to measure the mass of a sample first in the absence of a magnetic field, and then again when the magnetic field is switched on. The difference in mass can be used to calculate the *magnetic susceptibility* of the sample, and from the magnetic susceptibility the *magnetic moment* can be obtained. The magnetic susceptibility and thus the magnetic moment are due to *moving charges*. In an atom, the moving charge is due to electron transfer from one atomic orbital to another as we have studied in the freshman General Chemistry class. Similarly, color in transition metal complexes is due to an electron being excited from one d orbital to a higher energy d orbital. The energy difference for the first transition series generally falls in the visible spectral region. Absorption of one color in the visible spectrum results in the ion having the *complementary color*. The amount of d-orbital splitting depends on the nature of ligands in terms of its binding strength being high or low. As a result of this, different ligands have different splitting energies leading to different colors. Thus, bonding theories can explain color, magnetism, coordination number, and reactivity of *d*-block metal complexes.

Advanced Inorganic Chemistry. http://dx.doi.org/10.1016/B978-0-12-801982-5.00006-0

2. VALENCE BOND THEORY

The valence bond theory is an old theory that has not been extensively used, but some of its ideas and values can often be used to explain coordination chemistry.

2.1 Coordination compounds

Coordination compounds are Lewis acid–base complexes where the transition metal atom or ion acts as a Lewis acid and forms coordinate covalent bonds with the ligands.

A transition metal must make available a number of vacant orbitals equal to its coordination number to form the bonds. To get the most efficient overlap, the orbitals on the metal (s, p, or d orbital) will hybridize. The hybridizations and orbital geometries for the most common coordination numbers are (use group theory to show this):

1. Coordination number four

sp^3 or $sd^3(s, d_{xy}, d_{xz}, d_{yz}) \rightarrow$ tetrahedral

$dsp^2(s, p_x, p_y, d_{x^2-y^2}) \rightarrow$ square planar

2. Coordination number five

$dsp^3(d_{z^2}) \rightarrow$ trigonal bipyramid

3. Coordination number six

$d^2sp^3(d_{z^2}, d_{x^2-y^2}) \rightarrow$ octahedral

2.2 Coordination number six

1. Octahedral complexes are the most common among coordinated compounds of the *d*-block metals.

2. Several examples can be expected with the d^2sp^3 hybridization.

  a. Consider the electron configuration of free Co^{3+} ion in $Co(NH_3)_6^{3+}$ complex as:

$$\underset{3d}{\uparrow\downarrow \ \uparrow \ \uparrow \ \uparrow \ \uparrow} \qquad \underset{4s}{\_} \quad \underset{4p}{\_ \ \_ \ \_}$$

Spin pairing will free up two d orbitals to facilitate the d^2sp^3 hybridization as:

$$\underset{}{\uparrow\downarrow \ \uparrow\downarrow \ \uparrow\downarrow} \ \_ \ \_ \quad \_ \ \_ \ \_ \ \_$$

$$\downarrow d^2sp^3 \text{ hybridization}$$

$Co(NH_3)_6^{3+}$ $\underset{3d}{\uparrow\downarrow \ \uparrow\downarrow \ \uparrow\downarrow}$ $\underset{\underset{d^2sp^3 \text{ hybrids}}{NH_3 \ NH_3 \ NH_3 \ NH_3 \ NH_3 \ NH_3}}{xx \ \ xx \ \ xx \ \ xx \ \ xx \ \ xx}$

The NH_3 as a ligand is a strong enough base to cause the d electrons to pair up. Such complexes are called *low-spin, spin-paired,* or *inner orbital* complexes.

b. Another example is the hexafluorocobaltate(III) complex, $[CoF_6]^{3-}$. Fluoride anion (F^-) is not a strong enough base to cause spin pairing, so the metal must use its 4d orbitals in bonding the fluoride to the Co.

Free Co^{3+} $\quad$ ⇅ ↑ ↑ ↑ ↑ $\quad$ — $\quad$ — — — $\quad$ — — — — —
$\qquad\qquad\qquad$ 3d $\qquad\qquad$ 4s $\quad$ 4p $\qquad\qquad\qquad$ 4d

$\qquad\qquad\qquad\qquad\qquad$ $\Big\downarrow$ d^2sp^3 hybridization

$[CoF_6]^{3-}$ $\quad$ ⇅ ↑ ↑ ↑ ↑ $\quad$ xx $\;$ xx $\;$ xx $\;$ xx $\;$ xx $\;$ xx
$\qquad\qquad\qquad\qquad\qquad\quad$ F $\;\;$ F $\;\;$ F $\;\;$ F $\;\;$ F $\;\;$ F
$\qquad\qquad\qquad$ 3d $\qquad\qquad\qquad$ d^2sp^3 hybrids

This complex is called a *high-spin (spin-free, outer orbital)* complex.

c. In hexaamminenickel(II) $[Ni(NH_3)_6]^{2+}$, you can expect only the outer orbital complexation:

Free Ni^{2+} ⇅ ⇅ ⇅ ↑ ↑ $\quad$ — $\quad$ — — — $\quad$ — — — — —
$\qquad\qquad\quad$ 3d $\qquad\qquad$ 4s $\quad$ 4p $\qquad\qquad\qquad$ 4d

$\qquad\qquad\qquad\qquad\qquad$ $\Big\downarrow$ d^2sp^3 hybridization

$[Ni(NH_3)_6]^{2+}$ ⇅ ⇅ ⇅ ↑ ↑ $\quad$ xx $\quad$ xx $\;$ xx $\;$ xx $\quad$ xx $\quad$ xx
$\qquad\qquad\qquad\qquad\qquad\qquad$ NH_3 $\;$ NH_3 $\;$ NH_3 $\;$ NH_3 $\;$ NH_3 $\;$ NH_3
$\qquad\qquad\qquad$ 3d $\qquad\qquad\qquad$ d^2sp^3 hybrids

Thus, Ni^{2+} can form only *outer orbital complexes.*

2.3 Coordination number four

1. Square planar geometry is common for coordination number four as in the example tetracyanonickelate(II) $[Ni(CN)_4]^{2-}$:

Free Ni^{2+} ⇅ ⇅ ⇅ ↑ ↑ $\quad$ — $\quad$ — — — $\quad$ — — — — —
$\qquad\qquad\qquad$ 3d $\qquad\qquad$ 4s $\quad$ 4p $\qquad\qquad\qquad$ 4d

Spin pairing and hybridization yields:

$[Ni(CN)_4]^{2-}$ ⇅ ⇅ ⇅ ⇅ $\quad$ — $\quad$ — — — $\quad$ — —
$\qquad\qquad\qquad$ 3d $\qquad\qquad\quad$ 4s $\quad$ 4p

$\qquad\qquad\qquad\qquad\qquad$ $\Big\downarrow$ dsp^2 hybridization

$\qquad\qquad\qquad$ ⇅ ⇅ ⇅ ⇅ $\quad$ xx $\;$ xx $\;$ xx $\;$ xx $\quad$ —
$\qquad\qquad\qquad\qquad\qquad\qquad$ CN $\;$ CN $\;$ CN $\;$ CN
$\qquad\qquad\qquad$ 3d $\qquad\qquad$ dsp^2 hybrids $\qquad$ 4p

As a result of this, $[Ni(CN)_4]^{2-}$ and all square planar d^8 complexes are *diamagnetic*.

2. Tetrahedral complexes were observed for tetrachloronickelate(II)], $[NiCl_4]^{2-}$:

Free Ni^{2+} ⇅ ⇅ ⇅ ↑ ↑   _   _ _ _   _ _ _ _ _

           3d   4s   4p   4d

Spin pairing does not occur due to weak field Cl$^-$ ligand and, therefore, undergoes sp^3 hybridization:

$[NiCl_4]^{2-}$ ⇅ ⇅ ⇅ ↑ ↑   xx xx xx xx

                               Cl Cl Cl Cl

              3d   sp^3 hybrids

Consequently, $[NiCl_4]^{2-}$ complex and all tetrahedral d^8 complexes are *paramagnetic*.

3. CRYSTAL FIELD THEORY

The crystal field theory (CFT) considers the effects that the ligands will have on the energies of the d orbitals in a complex. To understand the CFT operation in complex formation, we must note the following:

1. Assume that the complex is held together by electrostatic interactions between the metal cation and the anionic or dipolar ligands.
2. Because we neglect the effects of covalent interactions, CFT is useful only in rationalizing the differences in properties of a series of compounds.
3. Review the orientations of d orbitals.

3.1 Octahedral complexes

The effect of $6L^-$ ligands on the energies of the d orbitals of an M^{2+} ion in an octahedral complex is such that the energies of all the d orbitals increase and their degeneracy is lifted as shown in Fig. 6.1. Apparently,

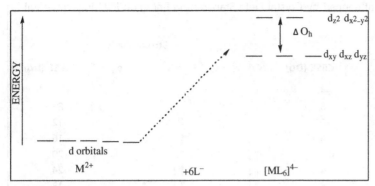

■ FIGURE 6.1 Effect of ligands on the energy of d orbitals of metal ion in an octahedral complex.

1. the $d_{x^2-y^2}$ and the d_{z^2} orbitals, which form the basis of the E_g irreducible representation, are higher in energy, because their *lobes of maximum probability point directly toward the ligands*. These two orbitals are called the *doubly degenerate e_g orbitals*;

2. the d_{xy}, d_{xz}, and d_{yz}, which form the basis of the T_{2g} irreducible representation, are lower in energy because their *lobes of maximum probability point away from the ligands*. These orbitals are called the *triply degenerate t_{2g} orbitals*, and

3. the difference in energy between the e_g and the t_{2g} orbitals (Δ_{Oh}) is called the *crystal field splitting energy*. Δ_{Oh} is often expressed in terms of an energy parameter called 10 Dq. Relative to the average energy of the d orbitals, each e_g orbital has an energy of +6 Dq and each t_{2g} orbital has an energy of -4 Dq. Preferential occupation of the lower energy t_{2g} orbitals will tend to stabilize the complex. This stabilization is called the *crystal field stabilization energy* (CFSE) and it is expressed in terms of Dq. However, distribution of electrons in these orbitals must obey the following rules (Table 6.1).

4. Follow Hund's rule and half-fill with spins parallel before pairing in a single orbital.

5. Preferentially fill the t_{2g} orbital before the e_g orbitals.

6. For the configuration of d^1 to d^3, t_{2g} should be singly occupied with spins parallel.

7. For the configuration of d^4 to d^7, there are two choices: (1) half-filled e_g before pairing in the t_{2g} to form *high-spin complexes*; (2) completely filled t_{2g} before filling the e_g to form *low-spin complexes*.

8. For the configuration of d^8 to d^{10}, there is only one possibility, that is, completely filled t_{2g}.

Table 6.1 Distribution of Electrons Influenced by Crystal Field Stabilization Energy (CFSE) in Octahedral Complexes

	Weak Field						Strong Field					
	t$_{2g}$			e$_g$		CFSE (Dq)	t$_{2g}$			e$_g$		CFSE (Dq)
D^1	↑					−4	↑					4
D^2	↑	↑				−8	↑	↑				−8
D^3	↑	↑	↑			−12	↑	↑	↑			−12
D^4	↑	↑	↑	↑		−6	↑↓	↑	↑			−16
D^5	↑	↑	↑	↑	↑	0	↑↓	↑↓	↑			−20
D^6	↑↓	↑	↑	↑	↑	−4	↑↓	↑↓	↑↓			−24
D^7	↑↓	↑↓	↑	↑	↑	−8	↑↓	↑↓	↑↓	↑		−18
D^8	↑↓	↑↓	↑↓	↑	↑	−12	↑↓	↑↓	↑↓	↑	↑	−12
D^9	↑↓	↑↓	↑↓	↑↓	↑	−6	↑↓	↑↓	↑↓	↑↓	↑	−6
D^{10}	↑↓	↑↓	↑↓	↑↓	↑↓	0	↑↓	↑↓	↑↓	↑↓	↑↓	0

3.1.1 Experimental evidence for crystal field stabilization

The heats of hydration (ΔH_{hyd}) for the high-spin first row M^{2+} ions are shown in Fig. 6.2 and in the following equation: $M^{2+}(g) + \infty H_2O \rightarrow M(H_2O)_6{}^{2+}(aq)$.

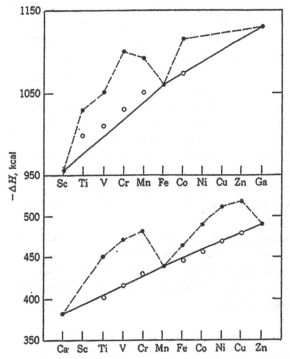

■ **FIGURE 6.2** The uncorrected (*dotted line*) and crystal field corrected (*solid line*) hydration energies of the divalent (*lower plot*) and trivalent (*upper plot*) ions of the first transition series. *Taken from Holmes, O. G.; McClure, D. S.,* J. Chem. Phys. **1957,** 26, 1686.

3.2 **Complexes of other geometries**

3.2.1 Tetrahedral complexes

In tetrahedral symmetry (Fig. 6.3), the four groups at alternate corners of the cube tetrahedrally coordinate with an atom at the center of the cube.

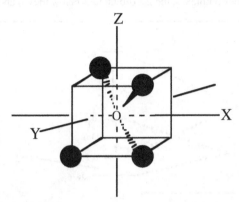

■ **FIGURE 6.3** Tetrahedral (cubic) symmetry.

If the central atom is a transition metal while the atoms in the corners are ligands, the metal d_{xy}, d_{xz}, and d_{yx} orbitals will be closer to the ligands than will the d_{z^2} and the $d_{x^2-y^2}$ orbitals.

The e_g orbitals will be the low-energy orbitals and the t_{2g} orbitals will be the high-energy ones. Since none of the orbitals point directly toward the ligands, the crystal field splitting energy is smaller than in octahedral complexes: $\Delta_{Td} = 4/9\Delta_{Oh}$ (Fig. 6.4).

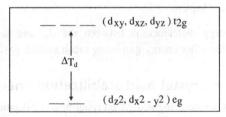

■ **FIGURE 6.4** The d orbital splitting in tetrahedral complexes.

Accordingly, each t_{2g} orbital is destabilized by an amount $2/5\Delta_{Td}$ and each e_g orbital is stabilized by $3/5\Delta_{Td}$.

3.2.2 Square planar complexes

Think of a square planar complex as resulting from the distortion of an octa-hedral complex by moving the ligands on the Z-axis away from the metal. Since this lowers ligand–ligand repulsion, the ligands in the XY plane will move in a little closer to modify the energy of the d orbitals as shown in Fig. 6.5. In some complexes, the d_{z^2} orbital falls below the d_{xz} and d_{yz} orbitals.

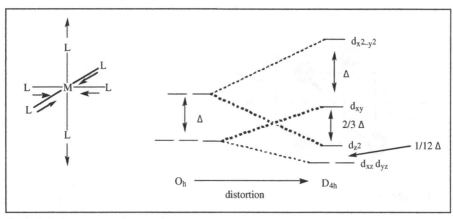

■ **FIGURE 6.5** The d orbital splitting in square planar complexes.

3.2.3 Complexes in a trigonal bipyramidal field

If we assume that the equatorial plane is the xy plane, we can obtain a qualitative picture of relative orbital energies. The ordering of the orbitals is:

d_{z^2} is the highest. Two ligands are on the z-axis.
d_{xy} and $d_{x^2-y^2}$ are intermediate.
d_{xz} and d_{yz} are lowest.

The largest energy difference is between the d_{z^2} and the d_{xy}, $d_{x^2-y^2}$ (≈ 8 Dq), with the other energy gap being much smaller (≈ 2 Dq).

3.3 Trends in crystal field stabilization energy (Δ)

1. General considerations regarding the crystal field stabilization energy (Δ):
   a. The values for Δ are of the order of about $40-210$ kJ mol^{-1}. Bond energies are of the order of $1200-6000$ kJ mol^{-1}. Therefore, Δ's are only about 2%–10% of bond energy.
   b. It cannot account for the general stability of a complex $\left(\Delta H_f^\circ\right)$, but may be able to account for the differences in the ΔH_f°'s of a series of compounds.

c. Experimentally, Δ is obtained from the visible/ultraviolet spectra of complexes and is often expressed in cm^{-1} ($= 1/\lambda$ in cm). This is a convenient spectroscopic energy unit.

$$1000 \text{ cm}^{-1} = 12.1 \text{ kJ/mol} = 2.9 \text{ kcal mol}^{-1}.$$

2. The Δ values for complexes of transition metals in the same period, with the same charge, and having the same ligands are of similar magnitudes. Example: Δ's for the high-spin $M(H_2O)_6^{2+}$ of the first row metals range from a low of 7500 cm^{-1} for $Mn(H_2O)_6^{2+}$ to a high of 14,000 cm^{-1} for $Cr(H_2O)_6^{2+}$.

3. The value of Δ increases as you go down a group and is from 30%– 50% per change in period (Table 6.2).

Table 6.2 Increase in Crystal Field Stabilization Energy (Δ) Values of the Complexes Using NH_3 Ligands With in the Trivalent Ion

Complex	Δ (cm^{-1})
$Co(NH_3)_6^{3+}$	23,000
$Rh(NH_3)_6^{3+}$	34,000
$Ir(NH_3)_6^{3+}$	41,000

4. For octahedral complexes in their normal oxidation states, the ligands arranged in the order of decreasing Δ are: $CO > CN^- > NO_2^- > $ bipy $>$ en $> NH_3 > CH_3CN > NCS^- > H_2O > C_2O_4^{2-} > OH^- > F^- > NO_3^- > Cl^- > SCN^- > S^{2-} > Br^- > I^-$.

Ligands with large Δ's have large back π bonding (CO, CN^-, NO_2^-, etc.), whereas H^-, NH_3, and CH_3^- ligands favor σ bonding. In general, $N > O$. However, Δ *is not related to the charge on the ligands*. It is an empirical series obtained spectroscopically.

5. Nonetheless, Δ *increases as the charge on the metal increases* and an increase in Δ was observed to be some 40%–80% in going from +2 to +3 complexes.

3.4 Predictions using crystal field theory: spin pairing of complexes

3.4.1 Octahedral complexes

1. High-spin/low-spin complexes are found for the d^4 through d^7 metals.
2. First row transition metal complexes will have low-spin complexes with ligands giving large Δ's (CN^-) and high-spin complexes with

ligands giving small Δ's (F^-). Thus, CoF_6^{3-} is a high-spin complex, whereas $Co(CN)_6^{3-}$ is a low-spin complex (Table 6.3).

Table 6.3 Mean Pairing Energies (cm^{-1}) for First Row Transition Metals		
Configuration	**M^{2+}**	**M^{3+}**
D^4	23,500	28,000
D^5	25,500	30,000
D^6	17,600	21,000
D^7	19,500	23,500

3. Since the pairing energy decreases as one goes down a group with increase in Δ's, in the second and third rows only low-spin complexes are found.

3.4.2 Tetrahedral complexes

Since $\Delta_{Td} \sim 4/9 \, \Delta_{Oh}$ and pairing energy does not change much, *only high-spin complexes* are possible.

3.4.3 Square planar complexes

Only high-spin complexes are observed for the d^1 to d^3 configurations. For the d^4 to d^8 configurations, either high-spin or low-spin complexes can be expected. Essentially all square planar d^7 and d^8 complexes are low-spin.

Ligands that have large Δ's favor square planar geometries in d^8 systems. Thus Pt(II), Pd(II), Au(III), Rh(I), and Ir(I) complexes are all diamagnetic, low-spin square planar complexes.

3.5 Distortions due to CFSE

If we assume that complexes have perfect octahedral or tetrahedral geometry, does this confirm the most stable configuration?

1. Let us consider an O_h Cu^{2+} (d^9) complex:

$$\underline{\uparrow\downarrow} \quad \underline{\uparrow} \qquad d_{z^2}, d_{x^2-y^2}$$

$$\underline{\uparrow\downarrow} \;\; \underline{\uparrow\downarrow} \;\; \underline{\uparrow\downarrow} \qquad d_{xy}, d_{xz}, d_{yz}$$

2. How does the asymmetric electron distribution in the e_g orbitals, that is, $(d_{z^2})^2$, $(d_{x^2-y^2})^1$, affect the overall geometry?

3. If the configuration is $(d_{z^2})^2$, $(d_{x^2-y^2})^1$, there will be less shielding of the nuclear charge in the xy plane and more shielding along the z-axis. This unequal shielding will cause the four ligands in the xy plane to be drawn in closer to the metal than the two ligands on the z-axis. Therefore there should be a distortion from O_h symmetry to D_{4h} symmetry. Since two electrons are stabilized while only one is destabilized, such a distortion will tend to stabilize the complex. The distortion due to an asymmetric electron distribution is called the *Jahn—Teller distortion* (Fig. 6.6).

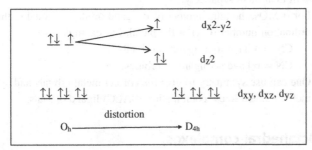

■ **FIGURE 6.6** The d orbital splitting in Jahn-Teller distortion.

Complexes having *asymmetric e_g electron distribution are all highly distorted*. Examples are (1) high-spin d^4 and d^9 and (2) low-spin d^7 complexes.

Complexes having *symmetric e_g and an asymmetric t_{2g} electron distribution are all slightly distorted*. Examples are (1) high-spin d^1, d^2, d^6, and d^7 and (2) low-spin d^4 and d^5.

Complexes having symmetric e_g and t_{2g} electron distributions are not distorted.

Examples are (1) high-spin d^3 and d^5 and (2) low-spin d^6 and d^{10}.

Tetrahedral complexes are not distorted.

4. MOLECULAR ORBITAL THEORY

In the molecular orbital (MO) theory approach, the MOs of the complex are constructed by taking a linear combination of ligand orbitals and metal atomic orbitals.

1. The metal can use its $(n-1)d$, ns, and np atomic orbitals.
2. There are two types of ligand orbitals that can be used and they are:
   σ orbitals

   a. Lobes point directly toward the metal.
   b. These are filled with electrons. They are the lone pair electron orbitals on the ligand.
   c. There is only one such orbital for each ligand.
π orbitals
   a. Lobes are perpendicular to the metal—ligand internuclear line.
   b. These could be either vacant or filled.
   c. There can be several such orbitals for each ligand.
3. There can be both σ and π MOs involving the metal and the ligands. We will treat them separately.

   For σ MOs, the total number of σ ligand orbitals is equal to the co-ordination number (CN) of the metal. For example,
   CN = 4 have four ligand σ orbitals.
   CN = 6 have six ligand σ orbitals.
   One can use symmetry to give the correct metal orbitals and sym-metry adapted linear combination (SALC) ligand orbitals.

4.1 Octahedral complexes

The σ ligand orbitals on the metal Cartesian axes are sketched in Fig. 6.7a. Each ligand could also have a set of π orbitals that could undergo π bonding with the metal. Therefore one can use a set of six vectors along the Cartesian axes to obtain the characters for the reducible representation for σ bonding and a set of 12 vectors for π bonding, as shown in Fig. 6.7b.

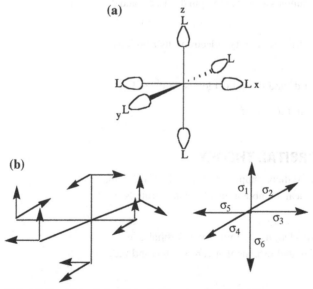

■ **FIGURE 6.7** (a) The σ ligand orbitals at *X*, *Y*, and *Z* coordinates. (b) A set of six vectors along the Cartesian axes.

Note that all 12 vectors for π bonding must be considered together since they can all be interchanged under the symmetry operations of the O_h point group (Table 6.4).

Table 6.4 Characters for the Reducible Representations

	E	$8C_3$	$6C_2$	$6C_4$	$3C_4^2$	i	$6S_4$	$8S_6$	$3\sigma_h$	$6\sigma_d$
Γ_σ	6	0	0	2	2	0	0	0	4	2
Γ_π	12	0	0	0	−4	0	0	0	0	0

Using the reduction equation we can obtain:

$$\Gamma_\sigma = A_{1g} + E_g + T_{1u} \qquad \Gamma_\pi = T_{1g} + T_{2g} + T_{1u} + T_{2u}$$

Referring to the O_h point group:

1. For σ bonding the metal can use its $s(A_{1g})$, $d_{z^2}, d_{x^2-y^2}(E_g)$; $p_x, p_y, p_z(T_{1u})$
2. For π bonding the metal can use its d_{xz}, d_{yz}, d_{xy} (T_{2g}); p_x, p_y, p_z (T_{1u}). Note that there are no metal orbitals with T_{1g}, or T_{2u} symmetry.

The atomic orbitals with the proper symmetry-adapted ligand σ orbitals leading to a qualitative MO correlation diagram for σ bonding is shown in Fig. 6.8. As seen in the correlation diagram, there are bonding A_{1g},

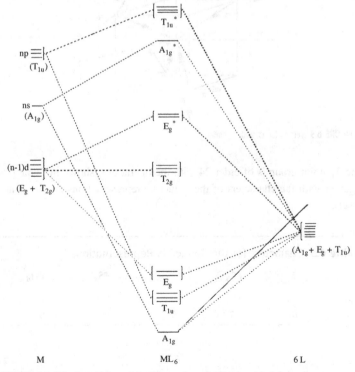

■ FIGURE 6.8 Molecular orbital correlation diagram for σ bonding in octahedral complexes.

T_{1u}, and E_g orbitals that are ligand based and can accommodate 12 electrons supplied by the ligands.

The next set of orbitals encountered are the nonbonding T_{2g} and antibonding E_g^* and they are metal based and will accommodate the metal electrons. Thus, CFT and MO theory give the same net result, that is, the metal electrons will be distributed in a triply degenerate t_{2g} set and a doubly degenerate e_g set, even though they differ in origin.

Δ = difference in energy between the nonbonding T_{2g} orbitals and the antibonding E_g^* orbitals. The generalizations about Δ, given in CFT, are also applicable to MO theory.

4.2 Other geometries

4.2.1 Tetrahedral complexes

A series of vectors for σ and π bonding is shown in Fig. 6.9. The molecule is oriented to take advantage of the cubic symmetry of a tetrahedron.

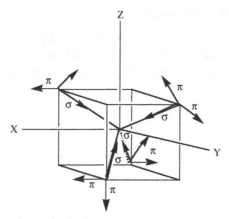

■ **FIGURE 6.9** Vectors for σ and π bonding.

The T_d point group is of order 24 with the operations, shown in Table 6.5, together with the characters of the reducible representations for the σ and π sets.

Table 6.5 Characters for the Reducible Representations

	E	$8C_3$	$3C_2$	$6S_4$	$6\sigma_d$
Γ_σ	4	1	0	0	2
Γ_π	8	−1	0	0	0

Using the reduction formula, $\Gamma_\alpha = A_1 + T_2$ $\Gamma_\pi = E + T_1 + T_2$.

For σ bonding the metal can use: $s(A_1)$ d_{xz}, d_{yz}, d_{xy} (T_2); p_x, p_y, p_z (T_2).

For π bonding the metal can use: $d_{z^2}, d_{x^2-y^2}$ (E), d_{xz}, d_{yz}, d_{xy} (T_2); $p_x, p_y,$ p_z (T_2).

The MO correlation diagram for σ bonding in T_d complexes, shown in Fig. 6.10, exhibits four low-energy, ligand-based bonding MOs that are

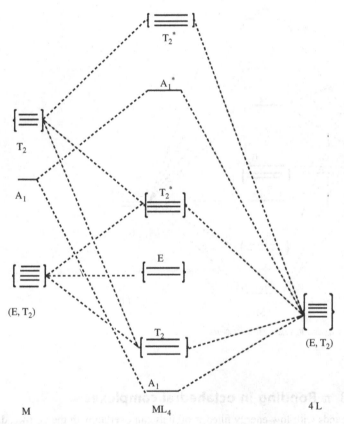

■ **FIGURE 6.10** Molecular orbital correlation diagram for σ bonding in T_d complexes.

filled with eight electrons (the "ligands electrons"). The metal d electrons will be distributed between a nonbonding doubly degenerate e_g set and a higher energy triply degenerate t_{2g}^* antibonding set of MOs. This is the same result as obtained by CFT.

4.2.2 Square planar complexes

The same distribution of metal d electrons as given by CFT is shown in Fig. 6.11.

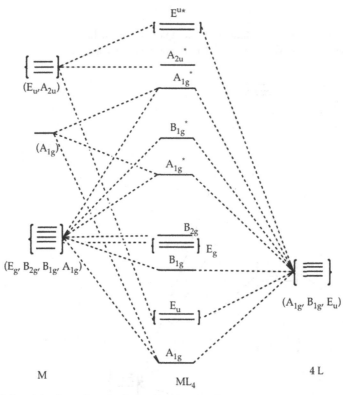

■ **FIGURE 6.11** Molecular orbital correlation diagram for σ bonding in square planar complexes.

4.3 π **Bonding in octahedral complexes**

Ligands with low-energy filled π orbitals can overlap with the t_{2g} (d_{xy}, d_{xz}, d_{yz}) metal orbitals (these orbitals were nonbonding in σ complexes). Therefore, the π MOs can be formed by taking a linear combination of the π ligand orbitals and the t_{2g} metal orbitals. In general, the ligand π orbitals will be lower in energy than the t_{2g} orbitals. Therefore, the low-energy bonding orbitals will be similar to the original ligand orbitals, whereas the high-energy orbitals will resemble the metal orbitals. This will give a correlation diagram as shown in Fig. 6.12.

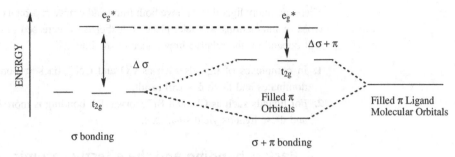

■ **FIGURE 6.12** Molecular orbital diagram for ligands with low-energy filled π orbitals.

The total number of electrons filling these MOs will be the ligand π orbitals, which will occupy the bonding π MOs, and the metal t_{2g} electrons, which are assigned to the higher energy π^* MOs. The net effect is to stabilize the complex with a decrease in Δ. Ligands such as OH^- and F^- give small Δ's.

For ligands with vacant high-energy π orbitals, π MOs can be formed by combining the ligand orbitals with the metal t_{2g} orbitals. Since the ligand orbitals, in general, have higher energies than the metal orbitals, the resulting low-energy π MOs will resemble the t_{2g} orbitals, and the high-energy π^* MOs will resemble the ligand orbitals. This gives rise to a correlation diagram as shown in Fig. 6.13.

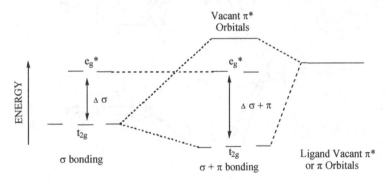

■ **FIGURE 6.13** Molecular orbital diagram for ligands with vacant high-energy π orbitals.

Since the ligand orbitals were vacant, the only π electrons will be those that occupied the metal t_{2g} orbitals. The interaction will lead to an increase in Δ. Therefore, ligands such as PH_3 and the olefins will produce large Δ's.

There are many ligands that have both filled and empty π orbitals, and both types of interactions will take place. Whether Δ is increased or decreased will depend on the relative importance of the interactions.

1. In complexes of ligands such as CO and CN⁻, back π bonding predominates and large Δ's will result.
2. For ligands such as Cl⁻ and Br⁻, forward π bonding is more important and these ligands yield small Δ's.

4.4 Back π bonding and the effective atomic number rule

In considering the formulas for organometallic compounds, a very convenient empirical rule is that the metals will tend to "complete" their valence shell, which for transition metals means that they would have a total electron count of 18. We can see the reason for this by considering the MO treatments for the simple metal carbonyls Fe(CO)$_5$ (Fig. 6.14) and Cr(CO)$_6$ (Fig. 6.15).

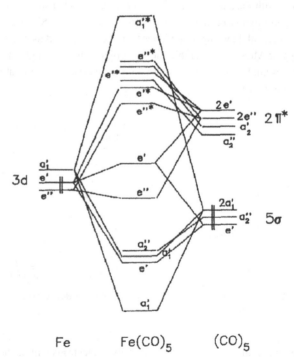

■ **FIGURE 6.14** Molecular orbital correlation diagram for Fe(CO)$_5$.

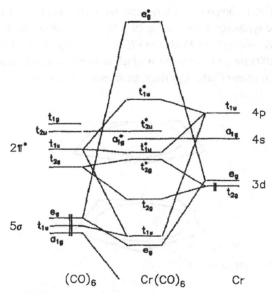

■ **FIGURE 6.15** Molecular orbital correlation diagram for $Cr(CO)_6$.

Fe(CO)$_5$: Fe(0) has eight valence electrons (d^6s^2) and it will bond with five CO's to attain its electron count of 18. Why is this so stable? Fig. 6.14 shows that the five CO σ electron pairs fill the four lowest energy, ligand-centered MOs to give a $(a_1')^2(e')^4(a_1')^2(a_2'')^2$ configuration. The eight electrons, "the Fe electrons", will occupy the two doubly degenerate, metal-centered MOs to give a $(e'')^4(e')^4$ configuration. Because of back π bonding, the energies of these two orbitals are lowered, while the energies of the antibonding pairs, $e''*$ and $e'*$ are raised. This results in a large energy gap, hence one would expect Fe(CO)$_5$ to be stable.

Cr(CO)$_6$: Cr(0) has six valence electrons (d^4s^2) and it will bond with six CO's to attain its electron count of 18.

The MO correlation diagram, Fig. 6.15, shows that six CO σ electrons will occupy the ligand-centered low-energy orbitals giving a $(a_{1g})^2(e_g)^4(t_{1u})^6$ configuration. The next six electrons will fill the metal-centered t_{2g} to give an overall $(a_{1g})^2(e_g)^4(t_{1u})^6(t_{2g})^6$ configuration. Because of back π bonding there is a large energy gap to the $t_{2g}*$ orbital. Therefore, Cr(CO)$_6$ should also be stable.

4.5 Arene complexes

One of the most important and interesting classes of organometallic compounds is the full- and half-sandwich metallocenes. A much studied example of such compounds is ferrocene, $(\eta^5\text{-}C_5H_5)_2Fe$, the first such compound to be synthesized (reported in 1951 by P. L. Paulson and T. K. Kealy and characterized by E. O. Fischer in 1952—see *J. Organomet. Chem.* Vols. *637–639* **2001** for retrospectives) in which an iron is situated symmetrically between two planar C_5H_5 (Cp) rings, as shown in Fig. 6.16.

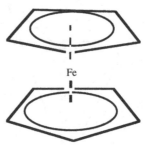

■ **FIGURE 6.16** Structure of ferrocene.

1. This is a stable orange compound that melts at 174°C and is stable to >500°C. The compound is synthesized by the reaction of Na or NaH with cyclopentadiene (C_5H_6) in tetrahydrofuran (THF) followed by the reaction of the resulting solution with $FeCl_2$. The equations are:

$$2C_5H_6 + 2Na \longrightarrow 2NaC_5H_5 + H_2$$

$$FeCl_2 + 2NaC_5H_5 \longrightarrow (C_5H_5)2Fe + 2NaCl$$

2. It is an open question as to how to view the compound, either as an Fe(0) bonded to two five-electron C_5H_5 radical fragments or as an Fe(II) bonded to two six-electron $[C_5H_5]^-$ ions. From the method of preparation, a formal +2 charge is usually assigned to the Fe. Within either approach, the two C_5H_5 (Cp) ligands are η^5-bonded to the Fe.

3. All evidence indicates that the Cp^- rings are aromatic and that the Fe metal orbitals interact with the filled π MOs on the open face of the Cp ligand. Fig. 6.1 shows the filled metal binding MOs for $[C_5H_5]^-$ and Fig. 6.17 shows the MO correlation diagram and sketches of the

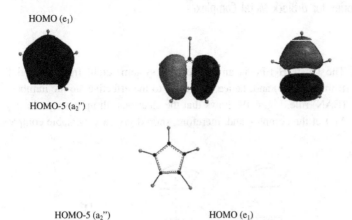

HOMO (e_1)

HOMO-5 (a_2'')

HOMO-5 (a_2'')   HOMO (e_1)

■ **FIGURE 6.17** Surfaces of the primary metal-binding orbitals viewed from above the ligands.

$[Cp_2]^{2-}$ and Fe^{2+} orbitals involved in bonding. These observations are typical for metallocenes.

The Cp_2^{2-} and symmetry compatible metal orbitals are shown in Fig. 6.18.

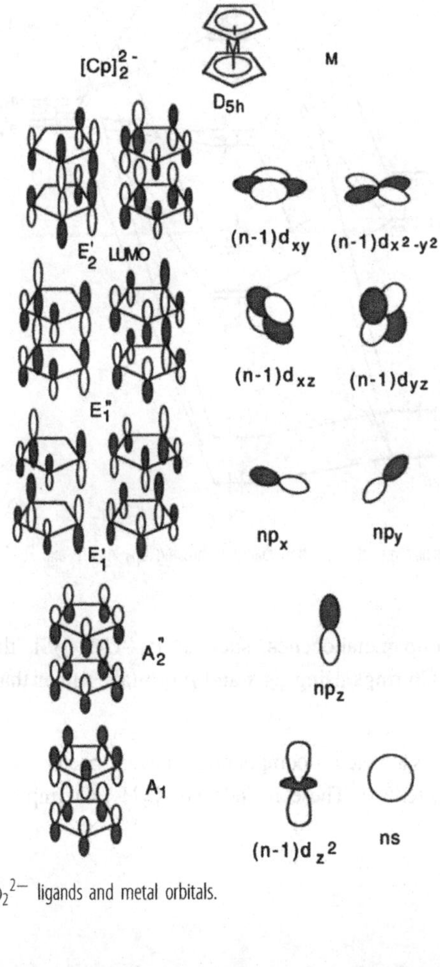

$[Cp]_2^{2-}$ M

D_{5h}

E_2' LUMO $(n-1)d_{xy}$ $(n-1)d_{x^2-y^2}$

E_1'' $(n-1)d_{xz}$ $(n-1)d_{yz}$

E_1' np_x np_y

A_2'' np_z

A_1' $(n-1)d_{z^2}$ ns

■ **FIGURE 6.18** Cp_2^{2-} ligands and metal orbitals.

4. The $(\eta^5-C_5H_5)_2Fe$ is an 18-electron system (eight from Fe and five from each Cp) and, hence, conforms to the effective atomic number (EAN) rule. Fig. 6.19 shows that the electrons fill up through the e_{2g} MO of the complex and, therefore, should give a very stable complex.

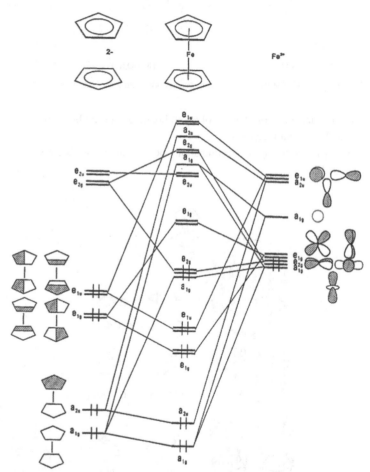

■ **FIGURE 6.19** Molecular orbital correlation diagram showing Cp_2 ligands and Fe^{2+} orbitals.

5. For main group metallocenes, such as $(\eta^5-C_5H_5)_2Si$, the metal atom bonds to the Cp rings using its *s* and *p* orbitals, rather than its d orbitals.

Other *d*-block sandwich compounds have been synthesized and structurally characterized. These include $(\eta^5-C_5H_5)_2Cr$ (mp = 168–170 °C),

$(\eta^5\text{-}C_5H_5)_2Ni$ (mp = 171–173 C), $[(\eta^5\text{-}C_5H_5)_2Co]^+$, and $(\eta^5\text{-}C_5H_5)_2TiCl_2$. In the last compound, the Ti, in a formal +4 state, is tetrahedrally surrounded by two η^5-bonding Cp rings and two Cl⁻ ligands (see structure II in Fig. 6.20). In general, many of the early transition metal sandwich compounds have this same structure, whereas the later ones tend to have a ferrocene-like structure.

In addition, both main group metals (and metalloids) and *f*-block metals form metallocenes. Both half- and full-sandwich complexes of the main group metals have been structurally investigated. Many of the full-sandwich metallocenes are bent (structure VI in Fig. 6.20), whereas in others the Cp rings are parallel to one another (structure V in Fig. 6.20). The $[(\eta^5\text{-}C_5Me_5)_2As]^+$ and $[(\eta^5\text{-}C_5Me_5)_2Sb]^+$ have bent structures, whereas $(\eta^5\text{-}C_5Me_5)_2Si$ crystallizes in a unit cell containing both parallel and bent silicocenes in a 2:1 ratio. The pentamethylcyclopentadienyl, $\eta^5\text{-}C_5Me_5$ (Cp*), is extensively used in place of the unsubstituted $\eta^5\text{-}C_5H_5$.

The lanthanides form complexes with up to three Cp (or Cp*) rings to form Cp_3Ln-type compounds (see Fig. 6.20:I). These compounds react with loss

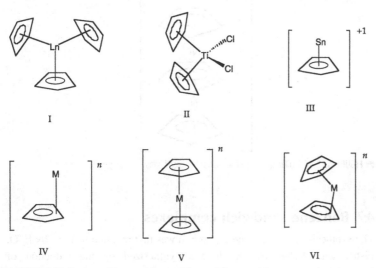

■ **FIGURE 6.20** Structures of some metallocenes (M = metal).

of a Cp to form bridged complexes of the type $[Cp_2Ln(\mu-X)_2LnCp_2]$. Many of the Cp_2Ln^{III} complexes can serve as promoters for the catalytic polymerization of olefins.

4.6 **Other arene-like ligands**

In addition to the Cp, other aromatic cyclic polyolefins have been used as ligands (Fig. 6.21).

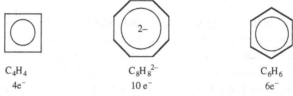

C_4H_4
$4e^-$

$C_8H_8^{2-}$
$10\,e^-$

C_6H_6
$6e^-$

■ **FIGURE 6.21** Structures of some cyclic polyolefins.

Examples include $(C_4H_4)Fe(CO)_3$, $CpTi(C_8H_8)$, $(C_6H_6)_2Cr$. It is also possible to form multidecker sandwich compounds such as $[(C_5H_5)_3Ni_2]^+$, whose structure is shown in Fig. 6.22.

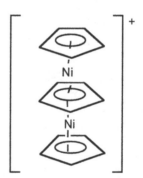

■ **FIGURE 6.22** Molecular structure of a multidecker nickel sandwich complex.

4.7 **Benzene sandwich complexes**

The synthesis of *bis*(benzene)chromium was first reported in 1955 by E. O. Fisher and Walter Hafner. It was synthesized by the reduction of $[(C_6H_6)_2Cr]^+$ by $S_2O_4^{2-}$ in basic solution according to the following equations.

$$6C_6H_6 + 3CrCl_3 + 2Al + xAlCl_3 \rightarrow 3\left[(C_6H_6)_2Cr\right][AlCl_4] \cdot (x-1)AlCl_3$$

$$2[(C_6H_6)2Cr]^+ + S_2O_4^{2-} + 4OH^- \rightarrow 2(C_6H_6)_2Cr + 2SO_3^{2-} + 2H_2O$$

The X-ray structure showed that the molecule has a D_{6h} symmetry and that each benzene was η^6-bonded to the Cr atom, as shown in Fig. 6.23.

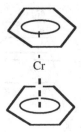

■ **FIGURE 6.23** Structure of $(\eta^6\text{-}C_6H_6)_2Cr$.

Chromium has also been found to form sandwich compounds with a number of different substituted benzenes and heterocycles. The orbital interactions for $(\eta^6\text{-}C_6H_6)_2Cr$ are shown in Fig. 6.24 and the MO correlation diagram is shown in Fig. 6.25.

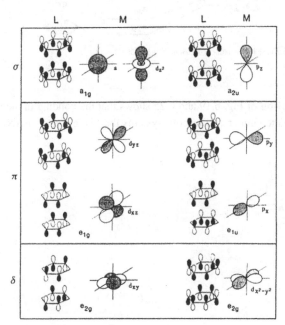

■ **FIGURE 6.24** Orbital interactions in $(\eta^6\text{-}C_6H_6)_2Cr$.

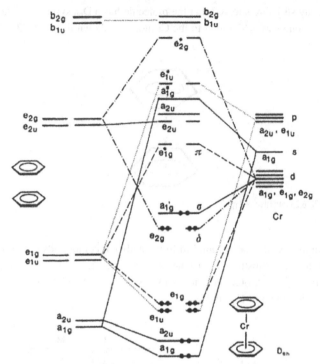

■ **FIGURE 6.25** Molecular orbital correlation diagram for $(\eta^6\text{-}C_6H_6)_2Cr$.

A comparison of Fig. 6.19 and 6.25 shows that the bonding in the two systems is quite similar and in both systems the "18 electron (EAN) rule" is obeyed.

Chapter 7

Coordination Chemistry: Reaction Mechanisms and Their Influencing Factors

1. INTRODUCTION: WHAT MAKES COORDINATION CHEMISTRY INTERESTING?

When transition metal salts are crystallized from water, the metal cations are closely associated with water molecules and they are called hydrated salts. Specifically, the water molecules surrounding these cations are examples of ligands. "**Ligand**" is a general term for a neutral molecule or ion (Lewis base) which is bonded to a metal. Metal—ligand bonding is often referred to as **coordination** of the metal with the ligand. The array of ligands is said to constitute the metal's **coordination sphere**. Finally, the entire ensemble of metal and ligands is called a **transition metal complex** or **coordination compound**. Why is it important to know more about these inorganic complexes? Researchers often emphasize that coordination compounds should not be overlooked in the context of biological implications, since their distinctive electronic, chemical, and photophysical properties are useful for variety of applications, such as metal-based NO sensors, metal complexes for phosphoprotein detection, protein labeling and probing DNA, as enzyme inhibitors, catalytic protein inactivators, magnetic resonance imaging contrast agents, and luminescent material for cellular imaging. These findings make bioinorganic chemistry a vibrant discipline at the interface of chemistry and the biological sciences, and also inspire chemists to synthesize new catalytic complexes to assist in a variety of chemical transformations.

Advanced Inorganic Chemistry. http://dx.doi.org/10.1016/B978-0-12-801982-5.00007-2

2. MODES OF SUBSTITUTION REACTION MECHANISMS

Generalized substitution reaction is:

$$M - X + Y \longrightarrow M - Y + X$$

where X is the ligand being replaced, Y is the incoming nucleophile, and M is the part of the metal complex that is not changed. However, there are two different modes of substitution.

2.1 Associative mode (a) (or an a intimate mechanism)

1. **Reaction coordinate**: $Y + M - X \longrightarrow [Y - M - X]^{\ddagger} \longrightarrow Y - M + X$
   a. The nucleophile will begin to form a bond with M, leading to an intermediate or transition state of increased coordination number.
   b. Bond making is more important than bond breaking. One aspect is how developed the Y—M bond is before M—X bond rupture begins.
2. **Intimate Mechanism—A or S_N2 (Limiting)**
   a. The Y—M bond is almost fully developed before the M—X bond begins to break. The Y—M—X species is an **intermediate** with a sufficient lifetime to be detected. A typical plot of energy (E) verses reaction coordinate is shown in Fig. 7.1.

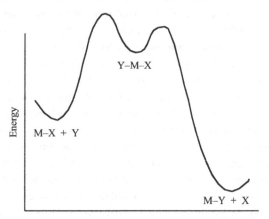

■ **FIGURE 7.1** Graphical illustration of intimate mechanism.

b. There are two transition states, the first in which bond making predominates, and the second in which bond breaking is more important. If X and Y were chemically identical, the two transition states would have the same energy. If X and Y are different, transition states of different energy would be expected, and either bond breaking or bond making could be rate limiting.

c. The mechanism would be:

$$M - X + Y \underset{k_{-1}}{\overset{k_1}{\rightleftharpoons}} YMX$$

$$YMX \xrightarrow{k_2} Y - M + X$$

Therefore, the rate law was determined to be: Rate $= \frac{k_1[M-X][Y]}{k_{-1}+k_2}$.

3. Interchange Associative Mechanism—I_a or S_N2

a. The generalized interchange mechanism is as follows:

$$M - X + Y \underset{k}{\overset{k}{\rightleftharpoons}} M - X, Y \quad \text{(fast equilibrium)}$$

$$M - X, Y \longrightarrow M - Y, X$$

$$M - Y, X \longrightarrow M - Y + X \quad \text{(fast)}$$

b. Rate law: Rate $= k[M-X,Y]$, but $[M-X,Y] = K[M-X][Y]$. Therefore Rate $= kK[M-X][Y]$.

c. The first step in this mechanism involves the nucleophile Y entering the second coordination sphere of the complex. If the species were both charged, this would be an ion pair. There is no chemical bond between Y and M.

d. While in the M—X,Y complex, a bond would begin to form between Y and M. At some point in the process, the M—X bond would begin to break. Both bond making and bond breaking would be important, but bond making would predominate.

e. An intermediate of increased coordination number cannot be detected; the species of higher coordination number could be a transition state. Thus this mechanism is called I_a (**interchange associative**) or S_N2.

2.2 Dissociative mode (d) (or a d intimate mechanism)

1. Intimate Mechanism—D or S_N1 (Limiting) can be illustrated as:

$$M - X \underset{k_{-1}}{\overset{k_1}{\rightleftharpoons}} M + X$$

$$M + Y \xrightarrow{k_2} MY$$

a. The rate law is determined to be: Rate $= \frac{k_1 k_2 [M-X][Y]}{k_{-1}[X] + k_2[Y]}$, and in the limiting process where $k_2[Y] \gg k_{-1}[X]$, the rate law can be written as Rate $= k_1[M-X]$. The rate is independent of the concentration and nature of Y.

b. In a D mechanism, M is an intermediate of reduced coordination number with a sufficient lifetime so that it can be detected.

c. Bond breaking is the only important thing with respect to the rate of reaction. Therefore, this mechanism is called **D (intimate mechanism)** or **S_N1 (limiting)**.

2. **Interchange Dissociative Mechanism (I_d) or S_N1**

a. This mechanism has the same general steps as the I_a mechanism, the difference being in the $M-X, Y \longrightarrow M-Y, X$ step, where bond breaking predominates.

b. A plot of energy verses the reaction coordinate is shown in Fig. 7.2.

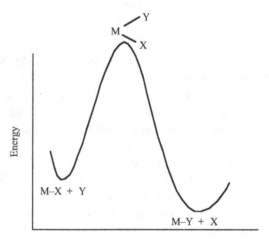

■ **FIGURE 7.2** Graphical illustration of an interchange dissociative mechanism.

The difference would be in the relative importance of the M−Y vs. the M−X bond in determining the energy of the transition state. Since the rate laws here are of limited use in determining the mechanism, other indirect methods must be employed.

3. COMPLICATIONS INVOLVING METAL COMPLEXES

3.1 Solvent competition

1. Many of the metal complexes are ionic and are soluble in water or some other polar solvent. Most polar solvent molecules are also nucleophilic.
2. The solvent can compete with the added nucleophile, Y, for the M—X complex. The reactions can go through an initial slow solvation step, followed by a rapid anation.

$$M - X + S \longrightarrow M - S + X \quad \text{slow}$$
$$M - S + Y \longrightarrow M - Y + S \quad \text{fast}$$

3. In such cases, the rate of substitution would be independent of the concentration and nature of the nucleophile irrespective of the mode of substitution.
4. If the solvent is nonpolar and nonnucleophilic, ion pair formation, as seen in the I mechanisms, could further complicate the straightforward interpretation of kinetic data.

3.2 Effects of changing the other ligands on the complex

One can make systematic changes in the complex and see how these affect the rate of the substitution reactions as shown in Table 7.1.

The kinetics results for the different charges can be considered in the light of Table 7.1.

4. ACTIVATION PARAMETERS

4.1 Enthalpy and entropy of activation

Consider a simple bimolecular reaction between A and B in terms of the Absolute Reaction Rate Theory. The reaction sequence is: Reactants $\longrightarrow$ Transition State $\longrightarrow$ Products

$$A + B \overset{K^{\ddagger}}{\rightleftharpoons} [A \cdots B]^{\ddagger} \longrightarrow C + D$$

$K^{\ddagger}$ is the pseudoequilibrium constant of the form: $K^{\ddagger} = \frac{[A \cdots B]^{\ddagger}}{[A][B]}$.

Table 7.1 Effects of Size and Charges on the Rates of **D** and **A** Reactions

Change	D and I_d	I_a	A
Increase positive charge of central atom	Decrease	Opposing effects	Increase
Increase size of the central atom	Increase	Increase	Increase
Increase the negative charge on the entering group	No effect	Increase	Increase
Increase the size of the entering group	No effect	Decrease	Decrease
Increase the negative charge on the leaving group	Decrease	Decrease	Decrease
Increase the size of the leaving group	Increase	Opposing effects	Decrease
Increase the negative charge of the other ligands	Increase	Opposing effects	Decrease
Increase the size of the other ligands	Increase	Decrease	Decrease

The activated complex breaks down to products through a vibration along the reaction coordinate and the rate can be estimated by the classical limit of vibrational frequency,

$$\nu = \frac{\kappa T}{h}, \left(\kappa = \text{Boltzman constant} \right.$$

$$= 1.381 \times 10^{-23} \text{ J/K and h is Planck's constant} = 6.626 \times 10^{-34} \text{ Js} \right)$$

The rate of the reaction would be Rate $= \frac{\kappa T}{h}[A \cdots B]^{\pm} = \frac{\kappa T}{h} K^{\pm}[A][B]$.

Therefore the specific rate constant, $k = \frac{\kappa T}{h} K^{\pm} = \frac{\kappa T}{h} \exp\left(-\frac{\Delta G^{\pm}}{RT}\right)$, and $\Delta G^{\pm} = \Delta H^{\pm} - T\Delta S^{\pm} (\Delta H^{\ddagger} = \text{enthalpy of activation and } \Delta S^{\ddagger} = \text{entropy of activation}).$

If both sides of the aforementioned equation are divided by T and the logarithm of both sides is taken, one obtains $\ln\left(\frac{k}{T}\right) = \ln\left(\frac{\kappa}{h}\right) + \frac{\Delta S^{\pm}}{R} - \frac{\Delta H^{\pm}}{RT}$

$\therefore$ a plot of $\ln\left(\frac{k}{T}\right)$ vs. $\frac{1}{T}$ should be a straight line, with a slope of $\frac{\Delta H^{\pm}}{R}$ and an intercept at $\frac{1}{T} = 0$ equal to $\ln\left(\frac{\kappa}{h}\right) + \frac{\Delta S^{\pm}}{R} = 23.76 + \frac{\Delta S^{\pm}}{R}$.

$\Delta H^{\ddagger}$ is related to the Arrhenius activation energy, E_a, by the relationship $\Delta H^{\ddagger} = E_a - RT$.

4.2 **Activation volume, $\Delta V^{\ddagger}$**

In the general sequence: Reactants $\longrightarrow$ Transition State $\longrightarrow$ Products, if there is a change in the molar volume in going from reactants to the transition state, then $\Delta V^{\ddagger} = V^{\ddagger} - V_R$. There should be a change in the specific rate constant with pressure, since:

$$\frac{\partial G}{\partial p} = V \quad \therefore \Delta V^{\pm} = -RT\frac{\partial \ln k}{\partial p}$$

This can be determined by using Le Chatelier's principle on the "equilibrium"

$$A + B \rightleftarrows [A\cdots B]^{\ddagger}$$

If $\Delta V^{\ddagger}$ is positive, an increase in pressure should inhibit the rate. On the other hand, if $\Delta V^{\ddagger}$ is negative, an increase in pressure should enhance the rate.

4.3 **Use of activation parameters in mechanistic studies**

1. In a dissociative mechanism (D or I_d) where bond breaking is important, you would expect the transition state to be larger and less structured than the product molecule. Therefore, one would expect that both $\Delta S^{\ddagger}$ and $\Delta V^{\ddagger}$ would be **positive**. It has been estimated that for a bimolecular reaction, $\Delta V^{\ddagger}$ would be about $+10$ cm^3/mol, and $\Delta H^{\ddagger}$ would be dependent on the M$-$X bond strength (X is the leaving group); it should be a fairly large number.
2. In an associative mechanism (A or I_a), the nucleophile$-$M bond is very developed in the transition state. Therefore, the transition state should be more ordered and more compact than that of the separated reactants. One would expect both a **negative** $\Delta S^{\ddagger}$ and $\Delta V^{\ddagger}$. It has been estimated that for an associative step, $\Delta V^{\ddagger}$ would be about -10 cm^3/mol.
3. Care must be taken when studying reactions in highly solvating solvents. If there is a large change in the number and orientation of solvent molecules in going from the reactants to the transition state, the activation parameters may be materially altered. In solvating solvents, it is convenient to think of an activation parameter, $\Delta X^{\ddagger}$, being the result of two contributions, an intrinsic one and a solvation contribution. That is $\Delta X^{\ddagger} = \Delta X^{\ddagger}_{intrinsic} + \Delta X^{\ddagger}_{solvation}$. In many instances, the

$\Delta X_{\text{solvation}}^{\ddagger}$ term would be of such a magnitude and sign to reverse the sign of $\Delta X^{\ddagger}$ from that expected for an associative or dissociative mechanism. Tables 7.2 and 7.3 illustrate the importance of solvation.

Table 7.2 Influence of Solvent on the Activation Volumes ($cm^3\ mol^{-1}$)

Negligible Solvent Effects		Significant Solvent Effects	
Dissociation			
$Cr(CO)_6 + PPh_3$	+15	$[Co(NH_5)_5Cl]^{2+}$ (aquation)	−9
$[Co(NH_3)(SO_3)]^+$ (NH_3 loss)	+6	$[Co(NH_3)_5(SO_4)]^+$ (aquation)	−19
$[Fe(phen)_3]^{2+}$ (aquation)	+15		
Bimolecular Reactions			
$W(CO)_6 + P(n\text{-Bu})_3$	−10	$[Co(NH_3)_5(N_3)]^{2+} + Fe^{2+}$ aq	+14
trans-$[IrCl(CO)(PPh_3)_2] + H_2$	−19	$[RhCl_6]^{3-} + Hg^{2+}$ aq	+22

Table 7.3 Activation Entropies for Some Bimolecular Reactions in Aqueous Solution

Reaction	$\Delta S^{\ddagger}$ ($J\ K^{-1}\ mol^{-1}$)	$Z_A Z_B$
$S_2O_3^{2-} + SO_3^{2-}$	−126	4+
$S_2O_3^{2-} + BrCH_2CO_2^-$	−71	2+
$ClCH_2CO_2^- + OH^-$	−50	1+
$S_2O_3^{2-} + BrCH_2COMe$	+25	0
$[Co(NH_3)_5Br]^{2+} + OH^-$	+92	2−
$ReCl_6^{2-} + Hg^{2+}$	+142	4−

Note that in the absence of solvation, the sign of $\Delta V^{\ddagger}$ is what is expected, whereas with solvation, sign reversals can occur. Also, it is apparent that the $\Delta S^{\ddagger}$'s for the bimolecular ionic reactions are more a function of the ion charges than anything else.

In reactions between like-charged ions where there is charge augmentation, the transition state is more ordered due to both a positive $\Delta S_{\text{intrinsic}}^{\ddagger}$ value and the fact that its higher charge would attract and orient more water molecules. Just the opposite would be expected if there was charge diminution in going from reactants to the transition state in which water molecules would be released, giving rise to a negative $\Delta S^{\ddagger}$ value.

5. EXAMPLES OF DIFFERENT COORDINATION NUMBERS WITH GEOMETRIES AND FACTORS INFLUENCING REACTION MECHANISM

The results will be discussed in terms of the reactions of the different coordination number for both main group and transition metal complexes.

5.1 Two- to six-coordinate complexes

1. **Two-coordinate complex**. Most are the d-block elements toward the right-hand end of the transition metals, notably the linear d^{10} complexes of Ag(I), Au(I), and Hg(II).

   a. All react very rapidly, many approaching the diffusion rate limit. All indications are that they react via an associative mechanism.

   b. Table 7.4 shows the second-order rate constants for the substitution of OH in MeHgOH. Note that the rate increases as the polarizability of the nucleophiles increases. This is expected for the associative mode of substitution at the soft Hg acid center.

Table 7.4 Second-Order Rate Constants for the Reaction: MeHgOH + X$^-$

X–	k (L mol^{-1} s^{-1})	X	k (L mol^{-1} s^{-1})	X	k (L mol^{-1} s^{-1})
Cl$^-$	1×10^4	Br$^-$	2×10^5	I$^-$	7×10^6
SCN$^-$	2×10^5	SO$_3^{2-}$	2×10^5		

   c. The fact that Br$^-$ substitution at EtHgOH is slower than at MeHgOH due to steric bulk is also consistent with an associative mode of substitution.

5.2 Three-coordinate complexes

This coordination is fairly rare; however, a number of compounds of p-block elements in Groups 13 and 15 have trigonal planar (group 13) or trigonal pyramidal (group 15) structures.

1. The ease and the propensity to form hypervalent compounds, at least for the heavier elements, should favor an associative mode of activation. Substitution at the trigonal planar borohalides, such as $R_2BCl + L \longrightarrow R_2BL^+ + Cl^-$, could proceed through one of four possible mechanisms: D (I_d), I_a, A, and, for nucleophiles with acidic hydrogens such as R'OH, a four-centered transition state of the form:

$$R_2B-Cl$$
$$\underset{R'}{\overset{}{\diagdown}}O-H$$

There is evidence for all of the possibilities.

   a. The orange $(R_2B)^+$ ion can be generated from $(Ph)_2BCl$ in nitrobenzene by the addition of the Lewis acid $AlCl_3$. This would make a D or I_d mode reasonable. However, efforts to trap the lower coordinate intermediate were not successful, so a D mechanism is unlikely.

   b. The rate constants are second order, but an analysis of the concentration–time curves did not indicate the existence of an R_2BClL intermediate, so a limiting A mechanism seems unlikely. Therefore, the most likely mechanism is an interchange (I_a or I_d). The following activation data has been collected for the reactions of BCl_3 and BBr_3 in acetonitrile (Table 7.5).

Table 7.5 Activation Parameters for $BX_3 + L \longrightarrow BX_2L^+ + X^-$

	$\Delta H^{\ddagger}$ (kJ mol^{-1})			$\Delta S^{\ddagger}$ (J K^{-1} mol^{-1})	
Base	BCl_3	BBr_3	Base	BCl_3	BBr_3
Dinitroaniline	40	80	Dinitroaniline	−77	+43
Dinitronaphthylamine	40	85	Dinitronaphthylamine	−65	+53

2. The negative $\Delta S^{\ddagger}$'s and moderate $\Delta H^{\ddagger}$'s for BCl_3 are consistent with an associative mode (I_a). The values for BBr_3 seem to indicate dissociative activation (I_d), or at least a process on the border between associative and dissociative (I).

   a. The hydrolysis of the phosphorus and nitrogen trihalides exhibit different products, indicating different substitution mechanisms:

$$NCl_3 + 3H_2O \longrightarrow NH_3 + 3HClO$$
$$PCl_3 + 3H_2O \longrightarrow HPO(OH)_2 + 3HCl$$

b. The ability of the P to form hypervalency allows for a low-energy nucleophilic water-O attack.

c. Since N does not form hypervalent compounds, a similar low-energy pathway is prevented in the NCl_3 reaction. The products are the result of a water-H attack on the N, accompanied by a bonding of the water-OH to the Cl.

d. The base hydrolysis of difluoramine, HNF_2, demonstrates another mechanism for substitution, the S_N1CB (unimolecular substitution of a conjugate base).

The rate law for the reaction $HNF_2 + 2OH^- \longrightarrow Products + H_2O$ can be written as:

$$Rate = -\frac{\partial[HNF_2]}{\partial t} = k[HNF_2][OH^-]$$

A number of different products are formed in the reaction including N_2F_2 and N_2, as well as N_2F_4, N_2O, NO_3^-, and NO_2^-. The problem is that, compared with other nucleophiles, the hydroxide reaction is uniquely fast, as can be seen in Table 7.6.

Table 7.6 Second-Order Rate Constants for Substitution at HNF_2

Base	k (L mol^{-1} s^{-1})	Base	k (L mol^{-1} s^{-1})
OH^-	6.9×10^2	Cl^-	5×10^{-4}
CN^-	4.6×10^{-2}	$CH_3CO_2^-$	4.7×10^{-5}
Br^-	8×10^{-3}	F^-	Very slow

The rate constant is approximately four orders of magnitude larger than the other rate constants, even though OH^- is not that much better of a nucleophile than the others. However, it is a good base.

e. This suggests the following mechanism:

$$HNF_2 + OH^- \underset{}{\overset{K}{\rightleftharpoons}} NF_2^- + H_2O \quad fast$$
$$NF_2^- \overset{k}{\longrightarrow} Products \quad slow$$

The rate law will be: $Rate = k[NO_2^-] = kK'[HNF_2][OH^-]$ where $K' = \frac{K}{[H_2O]}$.

f. It is an open question whether the aforementioned mechanism or a straightforward bimolecular displacement is the correct mechanism. This illustrates the complexity of trying to interpret kinetic data in a nucleophilic and highly solvating solvent such as water.

5.3 **Four-coordinate complexes**

There are numerous examples of four coordination in main group, d-block, and f-block elements. There are two geometries associated with this coordination: tetrahedral and square planar.

The square planar geometry is found in many of the d^8 complexes. Ni(II) with ligands producing large crystal field splittings and all Pt(II), Pd(II), Rh(I), Ir(I), and Au(III) four-coordinate complexes are square planar. The heavier group 18 $[YX_4]^-$ and XeX_4 structures are also square planar.

Tetrahedral geometries are one of the most common geometries, encompassing essentially all of the transition and main group elements.

5.3.1 Tetrahedral complexes of group 13

Because of the tendency of the tetrahedral B center to form three-coordinate complexes, it is not surprising that many substitution reactions at a tetrahedral B center proceed by a dissociative mechanism. For example, the rate of trimethylamine exchange in Me*BNMe$_3$ is independent of the concentration of NMe$_3$*, and has a $\Delta S^{\ddagger}$ of $+65$ J K^{-1}mol^{-1} and a $\Delta H^{\ddagger}$ that approximates the B$-$NMe$_3$ bond energy, indicative of a D mechanism.

In some cases, substitution proceeds through parallel dissociative and associative paths, as seen in Table 7.7 for the reaction of a number of substituted amine boranes with phosphines. The rate laws are all of the form: Rate $=$ $\{k_1 + k_2[L]\}$[amine borane]. Note that the k_{-1} data are consistent with a dissociative mechanism, and the data for the k_2 path are what one would expect for an associative mode of activation.

Table 7.7 Rate Constants for Substitution Reaction of Amine Boranes With Phosphines

Amine Borane	Nucleophile	k_1			k_2		
		$k_1 \times 10^5$	$\Delta H^{\ddagger}$	$\Delta S^{\ddagger}$	$k_2 \times 10^5$	$\Delta H^{\ddagger}$	$\Delta S^{\ddagger}$
n-BuH$_2$NMe$_3$	*n*-Bu$_3$P	0.24	159	$+167$	5.9	92	-38
i-BuH$_2$BNMe$_3$	*n*-Bu$_3$P	0.43	134	$+75$	3.2	109	-21
s-BuH$_2$NMe$_3$	*n*-BuP	1.3	130	$+79$	11.4	79	-29
t-BuH$_2$NMe$_3$	*n*-Bu$_3$P	11.5	134	$+105$	—	—	—

5.3.2 Tetrahedral complexes of group 14 (Si, Ge, Sn, and Pb)

All evidence indicates that the heavier group 14 tetrahedral complexes undergo substitution by way of an associative mode. The tendency to form hypervalent complexes suggests that a trigonal bipyramidal transition state of increased coordination should be quite accessible. For elements below Si, the reactions are very fast and are difficult to study.

The changes in stereochemistry are not as rigid as that found for carbon, in which inversion of configuration accompanies substitution via an associative mode. The reason is that the trigonal bipyramidal intermediate will have a sufficient lifetime such that the geometry can be scrambled by pseudorotation. Whether retention or inversion takes place seems to be dependent on the nature of the leaving group, the relative stabilities of the different ligand arrangements in the trigonal bipyramidal intermediate, and the polarity of the solvent.

5.4 Five-coordinate complexes of phosphorus and sulfur

The substitution reactions are all associatively activated. Much of the substitution chemistry of P can be explained on the basis of the formation of a five-coordinate intermediate of sufficient lifetime to undergo pseudorotation (Fig. 7.3).

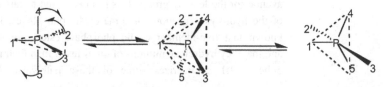

■ **FIGURE 7.3** Trigonal bipyramidal geometry of phosphorus.

Recall that as a result of the pseudorotation, one group (1) remains in the equatorial position. Also recall that the more electronegative ligands will tend to occupy the apical positions. This frequently controls the stereochemistry and course of many reactions.

The stereochemistry of the substitution reactions is not as straightforward as in C atom substitution where the five-member trigonal bipyramidal transition state leads to inversion of configuration (Walden inversion). The stereochemistry at the heavier group 14 and 15 centers depends greatly on the nature of the entering and leaving groups. There is a fine balance that dictates the stereochemistry of substitution; for example, halide substitution on $(t\text{-Bu})(\text{Ph})\text{SPX}$ (P = central atom; X = Cl or Br) by EtS^- proceeds with inversion when X = Cl, but with retention when X = Br, suggesting completely different mechanisms.

Table 7.8 Kinetic Data for $PhSO_2Cl + X$ Reactions (k at 25°C)

X	k (L mol^{-1} s^{-1})	$\Delta H^{\ddagger}$ (kJ mol^{-1})	$\Delta S^{\ddagger}$ (J K^{-1} mol^{-1})
OH$^-$	40.4	52.3	−38
PhNH$_2$	4.6	41.0	−94
C$_5$H$_5$N	3.1	51.0	−64
S$_2$O$_3$$^{2-}$	1.1	64.9	−26
F$^-$	0.66	61.9	−40

Table 7.8 lists the second-order rate constants and activation parameters for the substitution reactions of benzenesulfonyl chloride, $PhSO_2Cl$, with a number of nucleophiles (X) in aqueous solution. The parameters are consistent with an associative mode.

In general, the nucleophile is thought to approach and bond at a face of the tetrahedral reactant, leading to a transition state in which the nucleophile occupies an axial position; it is also thought that the group being displaced leaves from an axial position. This is reasonable in terms of the principle of microscopic reversibility. If the axial position provides the lowest energy path for the approach of a nucleophile, it also provides the lowest energy avenue for the leaving group. This has been used to rationalize the courses of the hydrolysis reactions of some cyclic phosphates. For example, it is known that five-member cyclic phosphate esters are sterically strained, and that they hydrolyze millions of times faster than their acyclic analogues. Scheme 7.1 summarizes some of these pathways. A five-coordinate

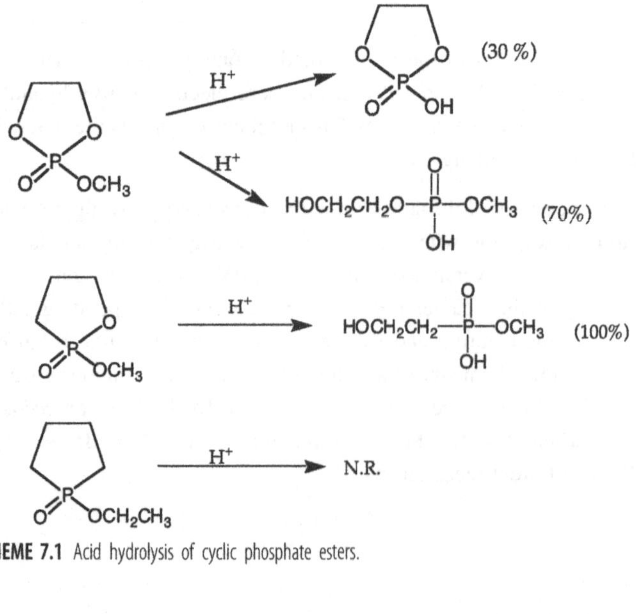

■ **SCHEME 7.1** Acid hydrolysis of cyclic phosphate esters.

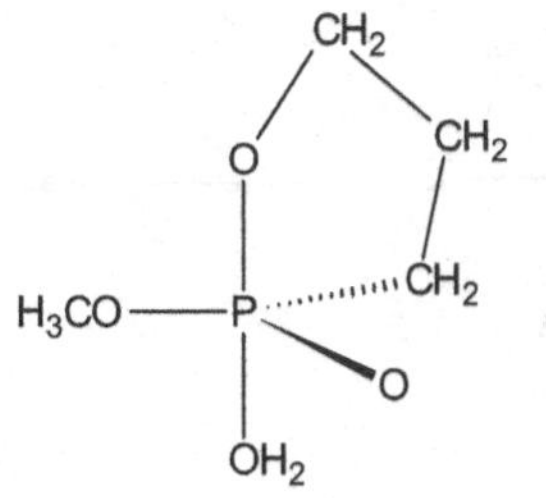

■ **FIGURE 7.4** A five-coordinate intermediate of a phosphate ester complexed with water.

intermediate can be postulated by the attack of water on the mono-oxy ester as shown in Fig. 7.4.

In this intermediate, there is no low-energy pseudorotational isomer that has the CH_3O in an axial position, so ring opening is the only mode. The same is true for the $(CH_2CH_2)_2PO(OC_2H_5)$ cyclic ester (third reaction in Scheme 7.1). Since the ring has no oxygen atom, ring opening is not possible and the compound is inert to hydrolysis.

5.5 **Square planar complexes**

The most studied are the d^8 complexes of Pt(II), Pd(II), Au(III), Rh(I), and Ir(I). All seem to react by the same mechanism. Of these, Pt(II) complexes are relatively inert and are the most studied, whereas the others are labile. Therefore, only the Pt(II) complexes will be discussed in any detail.

The rate law for the substitution reactions of these complexes is of the form:

$$MA_3X + Y \longrightarrow MA_3Y + X$$
$$\text{Rate} = k_1[MA_3X] + k_2[Y][MA_3X];$$

where MA_3X = the square planar complex, X is the leaving group, Y is the nucleophile, and A_3 stands for those ligands that are not replaced (they could be three different ligands).

1. All experimental results are consistent with an associative (**a**) mechanism for both the nucleophile-independent and the nucleophile-dependent paths.
2. The nucleophile-independent path involves a slow formation of the solvated complex, followed by rapid anation.

$$A_3M - X + S \longrightarrow A_3M - S \xrightarrow[\text{fast}]{+Y} A_3M - Y \quad (S = \text{solvent})$$

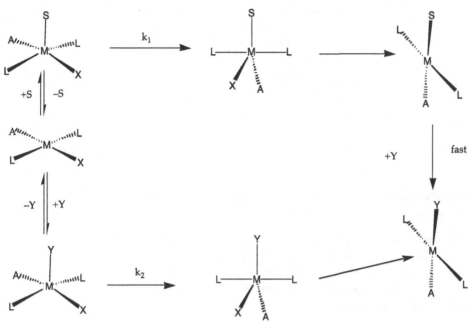

■ **SCHEME 7.2** Substitution reaction mechanism for square planar complexes.

3. The mechanism is shown in Scheme 7.2.

   Things to note are: (1) there is retention of configuration and (2) in the trigonal bipyramidal transition state, the nucleophile, the leaving group (X), and the ligand *trans* to the leaving group (A) are in the trigonal plane.

4. Some experimental data are given in Tables 7.9 and 7.10.

   Activation volumes for both the nucleophile-independent (k_1) and nucleophile-dependent (k_2) pathways are all negative.

Table 7.9 Acid Hydrolysis of Pt(II) Complexes at 20°C

Complex	$k_1 \times 10^5$ (s^{-1})	Complex	$k_1 \times 10^5$ (s^{-1})
$[PtCl_4]^{2-}$	3.9	*cis*-$[Pt(NH_3)_2Cl_2]$	2.5
$[Pt(NH_3)Cl_3]^-$	3.6	*trans*-$[Pt(NH_3)_2Cl_2]$	9.8
$[Pt(NH_3)_3Cl]^+$	2.6		

In noncoordinating solvents, the substitution reactions follow simple second-order rate laws. The rate of ligand exchange reaction

Table 7.10 Values of k_1 for Some Substituted Pt(II) Amine Complexes

Complex	Reagent	k_1 (s^{-1})
[Pt(dien)Cl]$^+$	Br$^-$	1×10^{-4}
[Pt(Et$_4$dien)Cl]$^+$	Br$^-$	8.5×10^{-6}
dien = H$_2$N–CH$_2$–CH$_2$–NH–CH$_2$–CH$_2$–NH$_2$;		
Et$_4$dien = (Et)$_2$N–CH$_2$–CH$_2$–NH–CH$_2$–CH$_2$–N(Et)$_2$.		

trans-[PdCl$_2$(Me$_2$S)$_2$] + *Me$_2$S $\longrightarrow$ [PdCl$_2$(Me$_2$S)(*Me$_2$S)] + Me$_2$S in dichloromethane is controlled by the equation:

Rate = k_2[PdCl$_2$(Me$_2$S)$_2$][Me$_2$S]

These data indicate that (1) the rate is fairly insensitive to the charge on the complex and (2) there is a steric inhibition to the reaction. The observations are all consistent with the (**a**) mechanism in which bond making predominates.

The k_2 increases as the polarizability of the nucleophile increases. This is consistent with the fact that these metals are soft acids and should react more readily with soft bases. The effect of the nucleophile on Pt(II) substitution can be expressed quantitatively using a parameter, n_{Pt}, which is defined by $n_{Pt} = \log(k_2/k_1)$ for the reaction in methanol:

$$\text{trans} - \left[\text{Pt(py)}_2\text{Cl}_2\right] + \text{Y}^{n-} \longrightarrow \text{trans} - \left[\text{Pt(py)}_2\text{ClY}\right]^{(n-1)-}$$
$$+ \text{Cl}^- \quad (\text{py} = \text{pyridine})$$

To make the ratio of rate constants dimensionally the same, the k_1 term was divided into $k_1 = k_1{}^0$[MeOH]. In pure methanol, [MeOH] = 24.3 mol/L. Therefore, the parameter usually used is $n_{Pt}{}^0 = \log(k_2/k_1{}^0)$. Table 7.11 lists some values of $n_{Pt}{}^0$ for the ligand substitution reaction: *trans*-[Pt(py)$_2$Cl$_2$] + Y^{n-} $\longrightarrow$ *trans*-[Pt(py)$_2$ClY]$^{(n-1)-}$ + Cl$^-$.

Table 7.11 Rate Constants and $n_{Pt}{}^0$ Values for Nucleophiles

Y	t (°C)	k_2 (M^{-1} s^{-1})	$n_{Pt}{}^0$	Y	t (°C)	k_2 (M^{-1} s^{-1})	$n_{Pt}{}^0$
CH$_3$OH	25	2.7×10^{-7}	0.00	MeO$^-$	25	Very slow	<2.4
Cl$^-$	30	4.5×10^{-4}	3.04	I$^-$	30	1.07×10^{-1}	5.46
CN$^-$	25	4.00	7.14	N$_3{}^-$	30	1.55×10^{-3}	3.58
NH$_3$	30	4.7×10^{-4}	3.07	PPh$_3$	25	249	8.93

a. In general, n_{Pt}^0 increases in the order: $I^- > Br^- > Cl^- >> F^-$; $Se > S > Te >> O$.

b. The reactions of Au(III) and Pd(II) complexes parallel those of Pt(II), but are much faster.

c. Although the activation mode is associative, it is not an **A** (S_N2 (lim)).

For the reaction: $[Pt(dien)X]^{(2-n)+} + Nu^{x-} \longrightarrow [Pt(dien)(Nu)]^{(2-x)+} + X^{n-}$ the rate constants for displacement decrease in the order $H_2O >> Cl^- > Br^- > I^- > N_3^-$.

5.6 **Trans effect**

One of the most striking aspects of the substitution reactions of Pt(II) complexes is the labilizing effect of the ligand *trans* to the leaving group in the complex and is called the **trans effect**. The ligands in **order of decreasing trans effect** are: $C_2H_4 \sim CN^- \sim CO \sim NO^+ \sim H^- > CH_3^- > SC(NH_2)_2 \sim PR_3 > NO_2^- \sim I^- \sim SCN^- > Br^- > Cl^- > NH_3 > OH^- > H_2O$.

In general, ligands that are good σ bonders (H^-, CH_3^-) or good back π bonders have high *trans* effects.

The labilizing effect can arise from interactions that either lower the energy of the transition state or increase the energy of the reactants. Depending on the ligands, both **labilizing and thermodynamic effects** are important.

The group being displaced (X) and the ligand *trans* to it (T) compete for the same orbital in bonding to the metal. Therefore, a good σ bonder should weaken the bond across from it. If one takes the p metal orbitals as an example, the change in possible bonding can be seen in Fig. 7.5. The

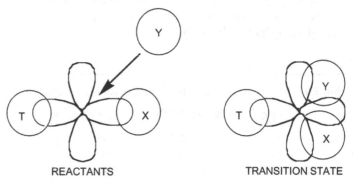

REACTANTS TRANSITION STATE

■ **FIGURE 7.5** Illustration of labilizing and thermodynamic effects.

incoming nucleophile, Y, begins to form a bond utilizing a vacant metal p orbital; in the transition state, both Y and X will use that orbital in bonding. Therefore, the transition state will not be as destabilized as the reactant by the T/X competition. There are some data on bond lengths to support this explanation. For example, the Pt–Cl bond distances in *trans*-[Pt(PR$_3$)$_2$TCl] for the different T's are: T = Cl$^-$ (2.294 Å); T = H$^-$ (2.422 Å); T = CH$_3$$^-$ (2.412 Å), and T = C$_2$F$_5$$^-$ (2.361 Å).

5.7 Kinetic effect

Good back bonding ligands, such as CN$^-$ and CO, operate in a different fashion. In the transition state, the leaving group (X), the nucleophile (Y), and the ligand *trans* to the leaving group (T) are in the trigonal plane (Fig. 7.6). A good back π bonding ligand will remove electron density

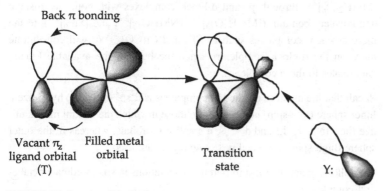

Back π bonding

Vacant π$_z$ Filled metal
ligand orbital   orbital
(T)

Transition
state

Y:

■ **FIGURE 7.6** Illustration of kinetic effects.

from the area opposite the metal where Y is approaching. This decrease in electron density along the reaction coordinate should lower the energy of the transition state, leading to increase in reaction rates.

5.8 Use of the trans effect in synthesis

The ligand Cl$^-$ has a greater *trans* effect than does NH$_3$, and therefore the *cis*- and *trans*- isomers of Pt(NH$_3$)$_2$Cl$_2$ can be synthesized by starting with either [PtCl$_4$]$^{2-}$ or [Pt(NH$_3$)$_4$]$^{2+}$ (Scheme 7.3).

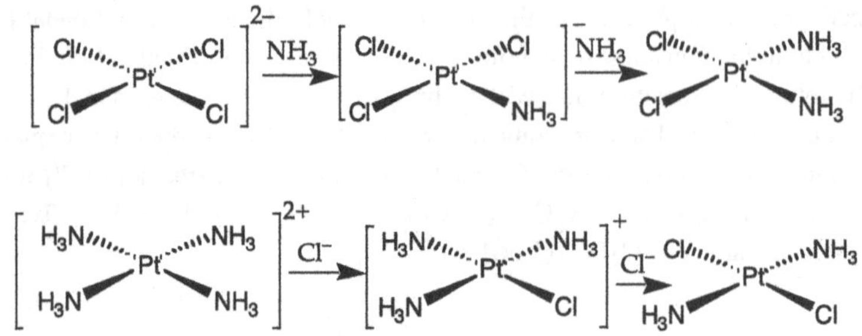

SCHEME 7.3 Syntheses of *cis-* and *trans-*isomers of Pt(NH$_3$)$_2$Cl$_2$.

5.9 Six-coordinate octahedral complexes

Six coordination and its associated octahedral geometry is by far the most commonly found arrangement in coordination chemistry. Examples range from *s*-block metals with predominately electrostatic bonding, such as [Mg(H$_2$O)$_6$]$^{2+}$, through p- and d-block complexes with both electrostatic and covalent bonding ([Mn(H$_2$O)$_6$]$^{2+}$, [Ni(H$_2$O)$_6$]$^{2+}$, and [SiF$_6$]$^{2-}$), to the more covalent complexes such as SF$_6$ and Cr(CO)$_6$. We will concentrate more on the d-block complexes, since the bulk of the available kinetic data relates to these complexes.

Recall that the metal in octahedral complexes is d^{2}sp^3 (or sp^{3}d^2) hybridized. Inner sphere (low-spin) complexes are those in which the central metals utilize their (n − 1) d$_{z^2}$ and d$_{x^2-y^2}$, n s and n p orbitals, whereas in the outer sphere (high spin) the metal n d$_{z^2}$ and d$_{x^2-y^2}$ are used.

The following are the most important observations in six-coordinate octahedral complexes.

1. In general, the outer sphere complexes are labile (Mn(II), Fe(II), Fe(III), Co(II), Zn(II), Cd(II), Hg(II), Al(III), Ga(III), In(III), and Tl(III)). For an isoelectronic series, lability decreases with increasing charge (oxidation number) on the central atom, e.g., AlF$_6^{3-}$ > SiF$_6^{2-}$ > PF$_6^-$ > SF$_6$.
2. The rates of substitution for inner sphere complexes depend on the number of d electrons that are on the metal, as can be seen in Table 7.12.
3. The aforementioned results are consistent with both associative and dissociative activation, so other experiments must be considered. Since many of the complexes are charged and are soluble in polar (nucleophilic) solvents, a straightforward interpretation of kinetic data is not possible. All evidence indicates that, in aqueous solution, substitution

Table 7.12 Inner Orbital Six-Coordinated Complexes

d Electron Configuration	Central Metal
Labile Complexes	
d^0 $(d^2sp^3)^{12}$	Sc(III), Y(III), Ln(III), Ti(IV), Zr(IV), Hf(IV) Ce(IV), Th(IV), Nb(IV), Ta(V), Mo(VI), W(VI)
d^1 $(d^2sp^3)^{12}$	Ti(III), V(IV), Mo(V), W(V), Re(VI)
d^2 $(d^2sp^3)^{12}$	Ti(II), V(III), Nb(III), Ta(III), Mo(IV), W(IV), Re(V), Ru(VI)
Inert Complexes	
d^3 $(d^2sp^3)^{12}$	V(II), Cr(III), Mo(III), W(III), Mn(IV), Re(IV)
d^4 $(d^2sp^3)^{12}$	$Cr(CN)_6^{4-}$, $Cr(bipy)_3^{2+}$, $Mn(CN)_6^{3-}$, Re(III), Ru(IV), Os(V)
d^5 $(d^2sp^3)^{12}$	$Cr(bipy)_3^{1+}$, $Mn(CN)_6^{4-}$, Re(II), $Fe(CN)_6^{3-}$, $Fe(Phen)_3^{3+}$ $Fe(bipy)_3^{3+}$, Ru(III), Ir(IV), Os(III)
d^6 $(d^2sp^3)^{12}$	$Mn(CN)_6^{5-}$, $Fe(CN)_6^{4-}$, $Fe(Phen)_3^{2+}$, $Fe(bipy)_3^{2+}$, Ru(II), Os(II) Co(III) (except CoF_6^{3-}), Rh(III), Ir(III), Ni(IV), Pd(IV), Pt(IV)

reactions go through a slow formation of an aquo complex, followed by rapid anation. Much data have been collected on solvation reactions.

4. For many substitution reactions of the type: $ML_6 + Y \longrightarrow$ Products, the reactions have been found to be independent of [Y] at high concentrations of Y, consistent with a dissociative mechanism, but at low concentrations, the rate depends on [Y], suggesting an associative mechanism. These seemingly conflicting results can be reconciled by the **Eigen–Wilkins mechanism**. This mechanism is based on the rapid formation of an **encounter complex**, which converts to products in a rate-determining step, as shown below.

$$\text{Step 1. } ML_6 + Y \underset{}{\overset{K_E}{\rightleftharpoons}} (ML_6, Y)$$
$$\text{Step 2. } (ML_6, Y) \overset{k}{\longrightarrow} \text{Products}$$

The complex (ML_6, Y) is an outer sphere complex that forms very rapidly; essentially, the formation of the complex is diffusion controlled. If both ML_6 and Y are charged, then (ML_6, Y) would be an ion pair. The equilibrium constant usually cannot be measured directly, but can be calculated.

$$K_E = \frac{[(ML_6, Y)]}{[ML_6][Y]}.$$

The (ML_6, Y) can rearrange in a slow step with Y entering the primary coordination sphere of the metal and changing places with an L. Whether the exchange is associative or dissociative depends on the relative importance of M$-$Y bond formation and M$-$L bond breaking in stabilizing the transition state. Therefore, the mechanism is either I_a or I_d.

$$\text{Rate} = k[(ML_6, Y)] = kK_E[ML_6][Y]$$

The total initial concentration of metal, $[M_{tot}]$, is measurable.

$$[M_{tot}] = [ML_6] + [(ML_6, Y)] = [ML_6] + K_E[ML_6][Y] = [ML_6]\{1 + K_E[Y]\}.$$

$$\text{Rate} = \frac{kK_E[M_{tot}][Y]}{1 + K_E[Y]}$$

Note that at low Y concentrations, $K_E[Y] << 1$ and the reaction is second order, first order in both ML_6 and Y. At higher concentrations the rate law is more complex, reaching a limit in which the rate is independent of $[Y]$.

Low concentration limit: $\text{Rate} = kK_E[M_{tot}][Y] = k_{obs}[M_{tot}][Y]$ and high concentration limit: $\text{Rate} = k[M_{tot}]$.

5. If the mode of substitution is dissociative, then k should be insensitive to the nature of Y. Since k_{obs} can be measured and K_E calculated, it is possible to estimate the values of k. For the reaction:
$Ni(H_2O)_6^{2+} + Y \longrightarrow Ni(H_2O)_5Y^{2+} + H_2O$, the values of k at 25°C are:

Y	NH$_3$	py	MeCO$_2^-$	F$^-$	SCN$^-$
$k \times 10^{-4}$, s^{-1}	3	3	3	0.8	0.6

These results are consistent with a dissociative mode of reaction.

6. One cannot generalize that a substitution in an octahedral complex goes exclusively by a dissociative path. This can be seen from the following activation volumes for water exchange (Table 7.13).

7. Examples of Co(III) amine complexes:
The Co(III) amines are the most studied of all the octahedral complexes; they can be easily synthesized and purified and are kinetically inert enough to be studied using classical methods. With the exception of OH$^-$ as a nucleophile (base hydrolysis),

Table 7.13 Activation Volumes ($cm^3\ mol^{-1}$) for Water Exchange Process

Ion[a]	$\Delta V^{\ddagger}$	Mode	Ion[a]	$\Delta V^{\ddagger}$	Mode
Al^{3+}	+6	I_d	Ln^{3+}	−6 to −7	I_a
Ga^{3+}	+5	I_d	$V^{2+}(d^3)$	−4	I_a
$Ti^{3+}(d^1)$	−12	I_a	$Mn^{2+}(d^5)$	−5	I_a
$V^{3+}(d^2)$	−9	I_a	$Fe^{2+}(d^6)$	+4	I_d
$Fe^{3+}(d^5)$	−5	I_a	$Co^{2+}(d^7)$	+6	I_d
$Ru^{3+}(d^5)$	−8	I_a	$Ni^{2+}(d^8)$	+7	I_d
$Rh^{3+}(d^6)$	−4	I_a	$Ru^{2+}(d^6)$	−0.4	I

Note that the results indicate the transition from an associative to a dissociative mode as the number of d electrons increases.
[a]All first-row transition metals are high spin.

the rates of substitution are all first order in the complex and zero order in nucleophile. All evidence indicates that the complexes undergo substitution by a rate-determining hydrolysis step, followed by rapid anation. Table 7.14 lists some rates of aquation

Table 7.14 Rates of Hydrolysis and K's for Co(III) Pentamines at 25°C

Y^{n-}	$k_h\ (s^{-1})$	K	Y^{n-}	k_h	K
NCS^-	3.7×10^{-10}	2.7×10^3	Br^-	3.9×10^{-6}	0.35
N_3^-	2.1×10^{-9}	8.3×10^2	I^-	8.3×10^{-6}	0.12
F^-	8.6×10^{-8}	2.5×10^1	NO_3^-	2.4×10^{-5}	0.08
Cl^-	1.8×10^{-6}	1.1			

(acid hydrolysis) for some pentamine cobalt(III) complexes. The reaction is:

$$[Co(NH_3)_5Y]^{(3-n)} + H_2O \overset{k_h}{\rightleftharpoons} [Co(NH_3)_5(H_2O)]^{3+} + Y^{n-}$$

where k_h is the rate constant for acid hydrolysis, and K is the equilibrium constant for the formation of the Co(III) amine complex.

Note that as the Co—Y bond strength increases, as measured by K for the complex formation, the rate of hydrolysis decreases. There is a linear relationship between log k_h and log K (Fig. 7.7).

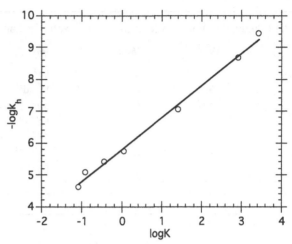

■ **FIGURE 7.7** Plot of $-\log k_h$ vs. $\log K$.

Note that $\Delta G^{\ddagger} \propto -\log k_h$ and $\Delta G^{\circ} \propto -\log K$. Therefore, a linear free energy relationship exists between $\Delta G^{\ddagger}$ and ΔG°. This indicates that the transition state is closely related to the products and that in the transition state, the Co—Y bond is very weak. These are all consistent with an I_d mechanism.

Table 7.15 Effect of Size of Nonsubstituted Ligands on the Rates of Acid Hydrolysis for $trans\text{-}[Co(N-N)_2Cl_2]^+ + H_2O \longrightarrow trans\text{-}[Co(N-N)_2Cl(H_2O)]^{2+} + Cl^-$

N—N	$k\ (s^{-1})$ at 25°C
$NH_2CH_2CH_2NH_2$	3.2×10^{-5}
$NH_2CH_2CH(CH_3)NH_2$	6.2×10^{-5}
$d,l\text{-}NH_2CH(CH_3)CH(CH_3)NH_2$	1.5×10^{-4}
$meso\text{-}NH_2CH(CH_3)CH(CH_3)NH_2$	4.2×10^{-4}
$NH_2C(CH_3)_2C(CH_3)_2NH_2$	3.2×10^{-2}

8. Steric effect

   The steric acceleration of rate, shown in Table 7.15, is indicative of an I_d mode.

9. Electronic effects

   The electronic effects on the rate in yielding $cis/trans$ isomers are shown in Table 7.16.

Table 7.16 Rates of Acid Hydrolysis at 25°C of the Reaction $cis/trans$-$[Co(en)_2LCl]^{n+}$ + $H_2O \longrightarrow$ $cis/trans$-$[Co(en)_2L(H_2O)]^{(n+1)+}$ + Cl^-

Cis isomer L	$k\ (s^{-1})$	% *Cis* in products	*Trans* isomer L	$k\ (s^{-1})$	% *Cis* in products
OH^-	0.012	84	OH^-	1.6×10^{-3}	75
Cl^-	2.4×10^{-4}	76	Cl^-	3.5×10^{-5}	26
NCS^-	1.1×10^{-5}	100	NCS^-	5.0×10^{-8}	60
NH_3	5×10^{-7}	100	NH_3	3.4×10^{-7}	0
H_2O	1.6×10^{-6}	–	H_2O	2.6×10^{-6}	–
CN^-	6.2×10^{-7}	100	CN^-	8.2×10^{-5}	0
NO_2^-	1.1×10^{-4}	100	NO_2^-	9.8×10^{-4}	0

Note that a good forward π bonding ligand, such as HO^- or Cl^-, causes acceleration of the rate, especially if they are in a *cis* position. This has been explained on the basis of L stabilizing the transition state by donating electron density to the orbital being vacated as the Cl^- leaves, as shown later. This is termed as the *cis* **effect** (Fig. 7.8).

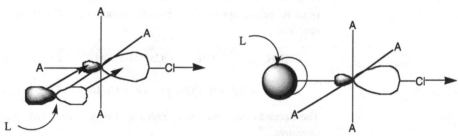

■ **FIGURE 7.8** *cis*-Effect during substitution.

If L is a good back π bonder, it is more effective in increasing the rate when it is in the *trans* position, presumably by stabilizing the five-coordinate transition state.

5.10 Isomerization during substitution

A tetrammine can exist as *cis* and *trans* isomers; the possibility of isomerization during substitution reactions can also be used as a test of mechanism.

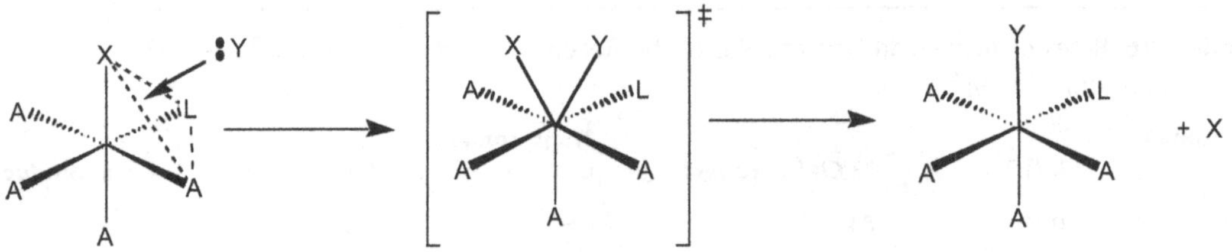

■ **SCHEME 7.4** Isomerization during associative mode of substitution.

In an associative mode of substitution, the incoming nucleophile will approach at a trigonal face containing the ligand to be substituted, X, to form a seven-coordinate transition of C_{2v} symmetry as shown in Scheme 7.4. Such a reaction will always lead to retention of configuration.

A dissociated mechanism could (not necessarily would) lead to isomerization, as shown in Scheme 7.5.

The results, shown in Table 7.16, are more consistent with a dissociative mechanism.

5.11 **Base hydrolysis**

As pointed out earlier, OH^- is the only nucleophile whose concentration appears in the rate law for the substitution reactions of Co(III) amines. For the reaction:

$$[Co(NH_3)_5Cl]^{2+} + OH^- \longrightarrow [Co(NH_3)_5(OH)]^{2+} + Cl^-, \text{ the rate law is :}$$

$$\text{Rate} = k[Co(III)][OH^-], \quad \text{where } Co(III) = [Co(NH_3)_5Cl]^{2+}.$$

The second-order rate could indicate that an associative mechanism is operative.

An alternative explanation is the dissociative conjugate base (D_{CB}) or S_N1-conjugate base (S_N1CB) mechanism (Scheme 7.6).

The first step involves the rapid removal of a proton on one of the coordinated ammines of the $[Co(NH_3)_5Cl]^{2+}$ complex by the OH^- anion to give the conjugate base $[Co(NH_3)_4NH_2]^+$ and H_2O. They are involved in the rapid equilibrium:

$$[Co(NH_3)_5Cl]^{2+} + OH^- \rightleftarrows [Co(NH_3)_4NH_2Cl]^+ + H_2O$$

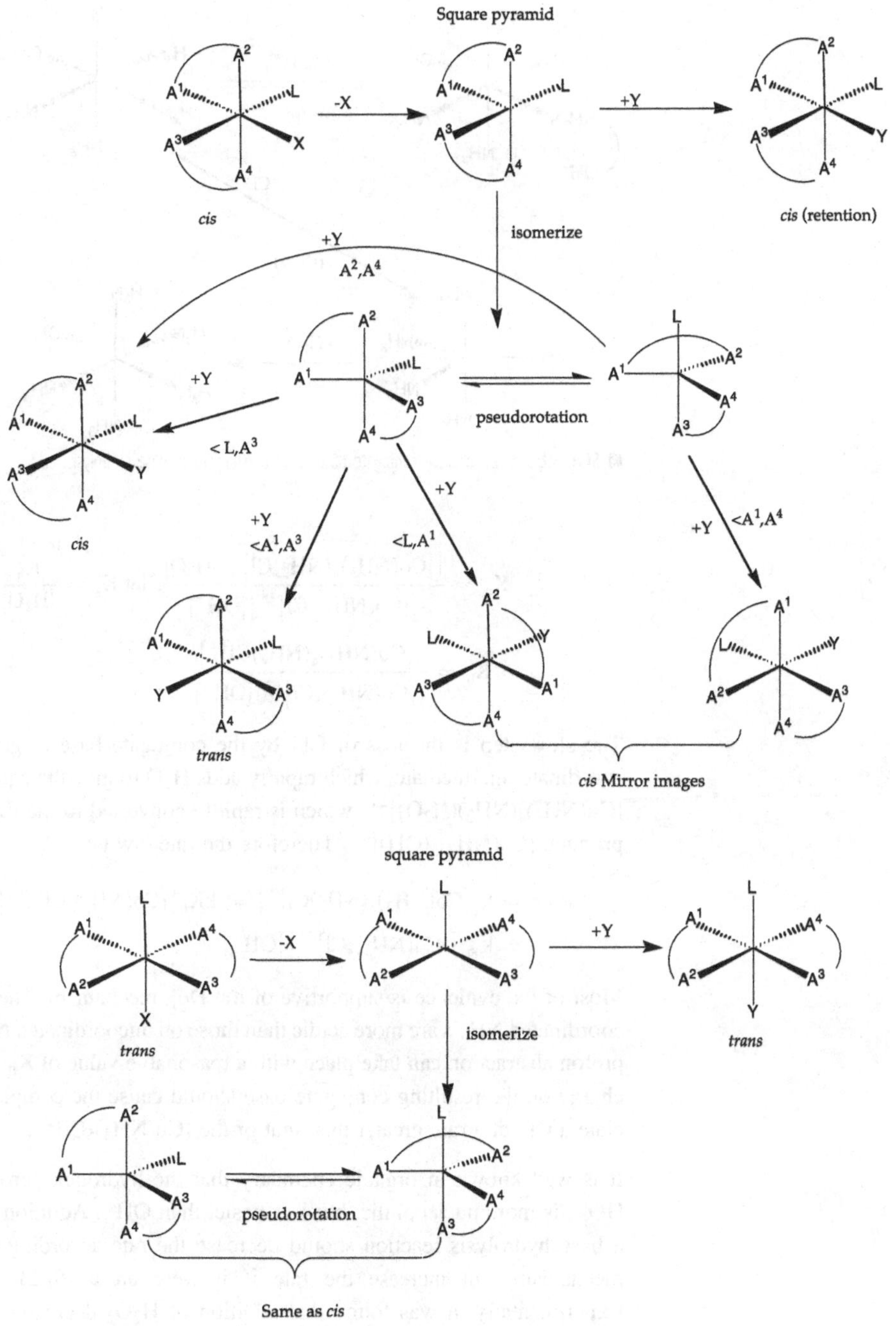

■ **SCHEME 7.5** Isomerization during dissociative mode of substitution.

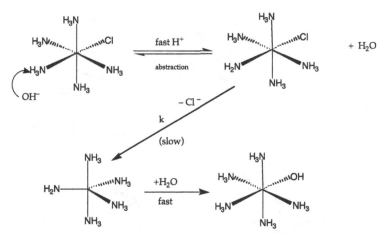

■ **SCHEME 7.6** Dissociative conjugate base (D_{CB}) mechanism in base hydrolysis.

$$K_h = \frac{\left[[Co(NH_3)_4(NH_2)Cl]^+\right][H_2O]}{\left[[Co(NH_3)_5Cl]^{2+}\right][OH^-]}, \text{ let } K_h' = \frac{K_h}{[H_2O]}$$

$$\therefore K_h' = \frac{\left[[Co(NH_3)_4(NH_2)Cl]^+\right]}{\left[[Co(NH_3)_5Cl]^{2+}\right][OH^-]}$$

The slow step is the loss of Cl^- by the conjugate base to give the five-coordinate intermediate, which rapidly adds H_2O to give the aquo complex, $[Co(NH_3)_4(NH_2)(H_2O)]^{2+}$, which is rapidly converted to the final hydroxo product, $[Co(NH_3)_5(OH)]^{2+}$. Therefore the rate law is:

$$\text{Rate} = k\left[[Co(NH_3)_4(NH_2)Cl]^+\right] = kK_h'\left[[Co(NH_3)_5Cl]^{2+}\right][OH^-]$$
$$= k_{obs}\left[[Co(NH_3)_5Cl]^{2+}\right][OH^-]$$

Most of the evidence is supportive of the D_{CB} mechanism. The H's on the coordinated NH_3's are more acidic than those on uncoordinated NH_3's so the proton abstraction can take place with a reasonable value of K_h'. The lower charge on the resulting conjugate base should cause the complex to dissociate a Cl^- at a rate greater than that of the $[Co(NH_3)_5Cl]^{2+}$.

It is well known in organic chemistry that the hydrogen peroxide anion, HO_2^-, is more nucleophilic, but less basic, than OH^-. Addition of H_2O_2 to a base hydrolysis reaction should decrease the rate according to the D_{CB} mechanism, but increase the rate if it were an **a** (S_N2) mechanism. Experimentally, it was found that addition of H_2O_2 decreased the rate of reaction.

If the reactions were via an associative mechanism, retention of configuration is always expected. Stereochemical analysis on the products of a number of base hydrolysis reactions confirmed the rearrangement of ligands. Examples: *trans*-$[Co(NH_3)_4(^{15}NH_3)Cl]^{2+}$ gave 60% *trans*- and 40% *cis*-$[Co(NH_3)_4(^{15}NH_3)(OH)]^{2+}$; *cis*-$[Co(en)_2(NH_3)Cl]^{2+}$ gave 22% *trans*- and 78% *cis*-$[Co(en)_2(NH_3)(OH)]^{2+}$; *trans*-$[Co(en)_2(NH_3)Cl]^{2+}$ gave 36% *trans*- and 64% *cis*-$[Co(en)_2(NH_3)(OH)]^{2+}$.

The D_{CB} mechanism requires the presence of a replaceable hydrogen. If the complex has no acidic hydrogen, base hydrolysis should be zero order in OH^-. The base hydrolyses of $[Co(py)_4Cl_2]^+$ and $[Co(CN)_5Cl]^-$ were found to be very slow and are zero order in OH^-. Although all evidence is consistent with a dissociative mechanism for the Co(III) amine complexes, it is not a D ($S_N1(lim)$).

1. Taube has studied the rates of acid hydrolysis of $[Co(NH_3)_5X]^{2+}$ (X = Cl, Br, I) in ^{18}O-enriched water in the presence of Hg^{2+}, Ag^+, and Tl^{3+} catalysts.
2. If the hydrolysis went by a D mechanism, each complex and catalyst should yield the same five-coordinate intermediate that should be able to discriminate between H_2O^{16} and H_2O^{18}. Because of the primary isotopic effect, the products should be enriched in H_2O^{18} by the same amount. The factor $f\left(\dfrac{[Co(NH_3)_5(H_2O^{18})]}{[Co(NH_3)_5(H_2O^{16})]} \cdot \dfrac{[H_2O^{16}]}{[H_2O^{18}]}\right)$ should be greater than one and the same for all complexes and catalysts. The results for Hg^{2+} catalyst (Table 7.17) show the same intermediate for each complex, presumably $[Co(NH_3)_5]^{3+}$, that can discriminate between H_2O^{18} and H_2O^{16}.

Table 7.17 *f* for $[Co(NH_3)_5X]^{2+}$

X	Catalyst		
	Hg^{2+}	Ag^+	Tl^{3+}
Cl	1.012	1.009	0.996
Br	1.012	1.007	0.993
I	1.012	1.010	1.003

The data for Ag^+ and Tl^{3+} catalysts indicated that H_2O bonding is also important in the transition state. Therefore, the mechanism is I_d.

5.12 **Other complexes**

We have discussed the substitution reactions of Co(III) in detail, not because it is typical for all octahedral complexes, but because much of the available data have been collected on this metal. Although the substitution reaction of Co(III) ion is via a dissociative mechanism, there are other examples of octahedral complexes that involve a different mechanism during substitution.

Most Cr(III) complexes seem to react via an associative mechanism (Table 7.18).

Table 7.18 Activation Volumes for Aquation of Some Cr(III) and Co(III) Pentaammine Complexes $[M(NH_3)_5L]^{3+} + H_2O \longrightarrow [M(NH_3)_5(H_2O)]^{3+} + L$

	Activation Volumes, $\Delta V^{\ddagger}$ (cm^3 mol^{-1})	
L	M = Cr^{III}	M = Co^{III}
H_2O	−5.8	+1.2
$OCHNH_2$	−4.8	+1.1
$OCHN(CH_3)_2$	−7.4	+2.6

5.12.1 Heavier group 9 metals

There is a marked decrease in reactivity as one proceeds down a group, as seen in Table 7.19. Since Dq increases as one goes down the group, the decrease in rate is consistent with either a **d** or an **a** mechanism.

Table 7.19 Rates of Chloride Exchange for *trans*-$[M(en)_2Cl_2]^+$ at 80°C

M	k (s^{-1})	$\Delta H^{\ddagger}$ (kJ mol^{-1})
Co	8×10^{-2}	98
Rh	2.5×10^{-5}	102
Ir	2.5×10^{-7}	106

Interpretation of the data leans more toward an associative mechanism, but much of the data seem to indicate an interchange substitution mechanism, as shown in Table 7.20.

Table 7.20 Activation Parameters for Solvent Exchange

Complex	k (s^{-1})	$\Delta H^{\ddagger}$ (kJ mol^{-1})	$\Delta V^{\ddagger}$ (cm^3 mol^{-1})	Substitution Mechanism
$[Rh(H_2O)_6]^{3+}$	2.2×10^{-9}	131	−4.2	I_a
$Rh(\eta^5\text{-}C_5Me_5)(H_2O)_3]^{2+}$	1.6×10^5	66	+0.6	$I_{(d)}$
$[Ru(H_2O)_6]^{2+}$	1.2×10^{-2}	88	−0.4	I
$[Ru(H_2O)_6]^{3+}$	−	90	−8.0	I_a
$[Ru(CH_3CN)_6]^{3+}$	8.9×10^{-11}	140	+0.4	I
$[Ir(H_2O)_6]^{3+}$	−	131	−6.0	I_a

5.13 Electron transfer reactions

In electron transfer reactions between transition metal complexes, the oxidizing agent and reducing agent must encounter each other for the transfer to take place. This leads to the question of whether ligands must be lost before the transfer.

Outer sphere electron transfer: No bonds between metal and ligand are broken, the oxidizing agent and the reducing agent are in secondary coordination spheres of each other.

Inner sphere electron transfer: A metal–ligand bond must be broken and a bimetallic species formed before electron transfer.

5.13.1 Inner sphere mechanism

Taube studied the reaction:

$$[Co(NH_3)_5Cl]^{2+} + [Cr(H_2O)_6]^{2+} + 5H_3O^+ \longrightarrow [Co(H_2O)_6]^{2+} + [Cr(H_2O)_5Cl]^{2+} + 5NH_4^+$$

Inert low spin   Labile high spin   Labile high spin   Inert low spin

In tracer studies, all Cr(III) ions were in the form of $[Cr(H_2O)_5Cl]^{2+}$ together with Cl labeled $[Co(NH_3)_5Cl]^{2+}$ in the presence of excess unlabeled Cl^-. It was shown that the Cl ligand in $[Cr(H_2O)_5Cl]^{2+}$ was the labeled one from the $[Co(NH_3)_5Cl]^{2+}$.

Since the Co center could not have lost the Cl before reduction, and the Cr center could not have gained it after oxidation, Cl transfer must have taken place while the redox process was taking place. This suggests a bimetalic intermediate such as in Fig. 7.9.

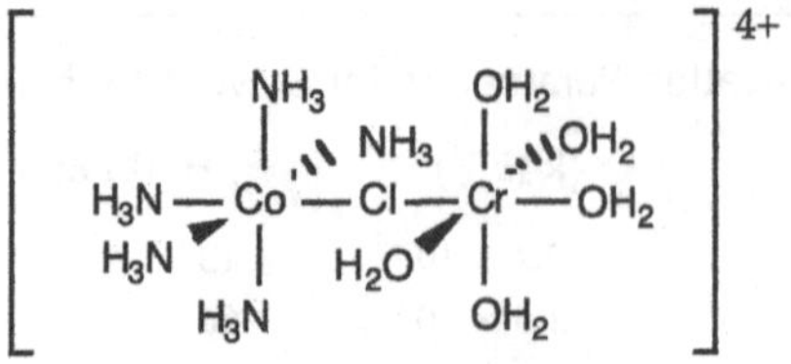

■ **FIGURE 7.9** Bimetallic intermediate in inner sphere substitution mechanism.

Common bridging ligands are halides, $C\equiv N^-$, OH^-, SCN^-, and heterocycles such as those shown in Fig. 7.10.

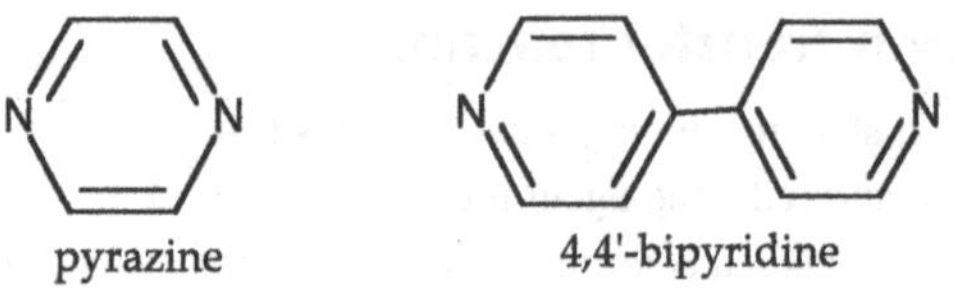

pyrazine   4,4'-bipyridine

■ **FIGURE 7.10** Common heterocyclic ligands.

The reaction can be broken down into several steps:

$$[Co(NH_3)_5Cl]^{2+} + [Cr(H_2O)_6]^{2+} \underset{k_{-1}}{\overset{k_1}{\rightleftharpoons}} [(NH_3)_5Co^{III}(\mu - Cl)Cr^{II}(H_2O)_5]^{4+} + H_2O$$

$$[(NH_3)_5Co^{III}(\mu - Cl)Cr^{II}(H_2O)_5]^{4+} \underset{k_{-2}}{\overset{k_2}{\rightleftharpoons}} [(NH_3)_5Co^{II}(\mu - Cl)Cr^{III}(H_2O)_5]^{4+}$$

$$[(NH_3)_5Co^{II}(\mu - Cl)Cr^{III}(H_2O)_5]^{4+} \underset{k_{-3}}{\overset{k_3}{\rightleftharpoons}} [Co(NH_3)_5]^{2+} + [Cr(H_2O)_5Cl]^{2+}$$

Most inner sphere complex reactions follow second-order kinetics, but the results are difficult to interpret. For some reactions, formation of the bridged complex could be the rate-determining step, the electron transfer step in others might be the slow step, or the dissociation of the bridge.

In the earlier reaction, the first and the last steps are fast (both $[Cr^{II}(H_2O)_6]^{2+}$ and $[Co^{II}(NH_3)_5]^{2+}$ are labile), which means that electron transfer is limiting. The rate should be sensitive to the nature of the bridging ligand, and this is found experimentally. In this step, although no metal−ligand bonds are broken, there must be an adjustment of the metal−ligand bonds. For example, in going from the d^4 Cr(II) to the d^3 Cr(III), one goes from a large distorted complex to a smaller symmetrical one. Structural adjustments must also accompany the Co(III) to Co(II) reduction.

If the d^3 $[V(H_2O)_6]^{2+}$ is substituted for the d^4 $[Cr(H_2O)_6]^{2+}$, then the formation of the bridged complex becomes the rate-determining step. The rate of electron transfer is also similar to the rate of water exchange for $[V(H_2O)_6]^{2+}$. Nonetheless, the rate does not vary much when a different bridging ligand is present on the Co(III) ammine complex.

5.13.2 Outer sphere mechanisms

When both the oxidant and reductant are inert, at least compared with the rate of electron transfer, electron transfer must take place by a tunneling or an **outer sphere mechanism**.

Examples are

$$\left[{}^*Co(NH_3)_6\right]^{2+} + \left[Co(NH_3)_6\right]^{3+} \longrightarrow \left[{}^*Co(NH_3)_6\right]^{3+} + \left[Co(NH_3)_6\right]^{2+}$$
$$\left[Fe(CN)_6\right]^{4-} + [IrCl_6]^{2-} \longrightarrow \left[Fe(CN)_6\right]^{3-} + [IrCl_6]^{3-}$$

The first reaction is an example of a **self-exchange** reaction in which there is no net chemical change ($\Delta G^0 = 0$), whereas the second is referred to as a **cross-reaction**. The general steps in such a reaction are those given in Scheme 7.7.

The reactants must approach close enough together for the electron to be transferred. An important principle that governs this process is known as the **Franck–Condon principle** (a molecular electronic transition is much faster than molecular motion). Therefore, during electron transfer, the atoms are essentially motionless so that the electron transfer can be only between vibrationally excited states with identical structures. This is termed an **encounter complex,** as depicted in Scheme 7.7.

The greater the changes in bond lengths required to reach the encounter complex, the slower the reaction. Thus, the self-exchange between $[Fe(CN)_6]^{4-}$ (low-spin) and $[Fe(CN)_6]^{3-}$ (low-spin) is $\sim 10^5 \, L \, mol^{-1} \, s^{-1}$ compared with $[Co(NH_3)_6]^{2+}$ (high-spin) and $[Co(NH_3)_6]^{3+}$ (low-spin), which is $\sim 10^{-6} \, L \, mol^{-1} \, s^{-1}$.

The more the ligand can facilitate electron transfer, the faster the electron transfer rate. For example, the rate for $[Co(phen)_3]^{3+}$ and $[Co(Phen)_3]^{2+}$ is $\sim 40 \, L \, mol^{-1} \, s^{-1}$ compared with $\sim 10^{-6} \, L \, mol^{-1} \, s^{-1}$ for the amine complex.

The various contributions to $\Delta G^{\ddagger}$ are represented by:

$$\Delta G^{\ddagger} = \Delta_w G^{\ddagger} + \Delta_o G^{\ddagger} + \Delta_s G^{\ddagger} + RT \ln \frac{\kappa T}{hZ}.$$

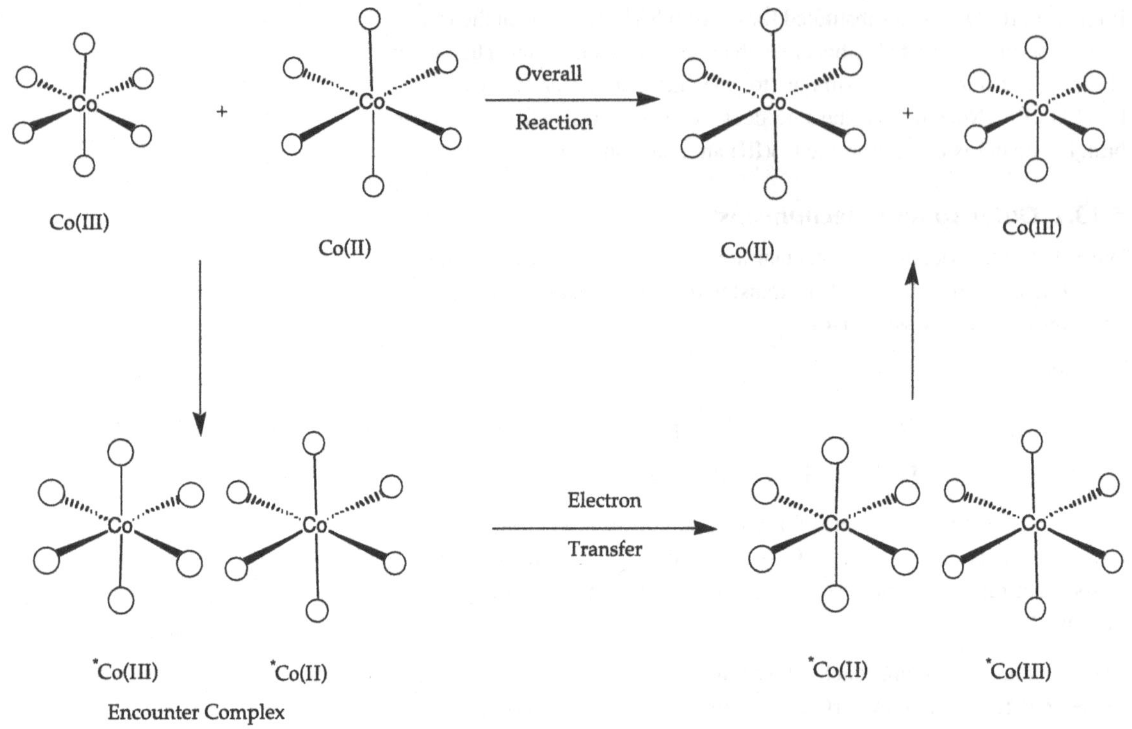

SCHEME 7.7 Formation of encounter complex via outer sphere mechanism.

$\Delta_W G^{\ddagger}$ is the contribution associated with bringing the oxidizing agent and the reducing agent together and includes the work done in overcoming electrostatic repulsions.

$\Delta_O G^{\ddagger}$ is the contribution due to changing bond lengths.

$\Delta_S G^{\ddagger}$ is the contribution arising from the rearrangement of solvent spheres.

The $RT \ln \frac{\kappa T}{hZ}$ term accounts for the loss of translational energy and rotational energy in forming the encounter complex ($\kappa =$ Boltzmann constant, $h =$ Planck's constant, and Z is the collision number $\sim 10^{11}$ L mol^{-1} s^{-1}).

The values of each contribution can be calculated from the **Marcus—Hush** theory, and $\Delta G^{\ddagger}$ can be calculated for self-exchange reactions.

The Marcus—Hush theory can be used to calculate the rate constant for a cross-exchange reaction (k_{12}) from the rate constants for the two self-exchange reactions (k_{11} and k_{22}) and the equilibrium constant of the cross-exchange reaction (K_{12}).

According to the Marcus theory:

$$k_{12} = (k_{11}k_{22}K_{12}f_{12})^{1/2} \text{ where } f_{12} \text{ is defined by the relationship, } \log f_{12}$$

$$= \frac{(\log K_{12})^2}{4 \log\left(\dfrac{k_{11}k_{22}}{Z^2}\right)}$$

The f_{12} term corrects for the effect of the differences in free energies of the two reactants, but for most reactions, $f_{12} \sim 1$.

The equation is very successful, and when the rate of a cross-exchange reaction can be calculated by the Marcus theory, it is taken as proof of an outer sphere mechanism.

Example: The equilibrium constant for the following cross-exchange reaction is 2.6×10^5 and its rate constant is $1.5 \times 10^4 \, \text{L mol}^{-1} \, \text{s}^{-1}$.

$$\left[Ru(NH_3)_6\right]^{2+} + \left[Co(phen)_3\right]^{3+} \longrightarrow \left[Ru(NH_3)_6\right]^{3+} + \left[Co(phen)_3\right]^{2+}$$

Given the rate constant for the $[Co(phen)_3]^{3+}/[Co(phen)_3]^{2+}$ exchange is $8.2 \times 10^2 \, \text{L mol}^{-1} \, \text{s}^{-1}$ and the $[Ru(NH_3)_6]^{3+}/[Ru(NH_3)_6]^{2+}$ exchange is $40 \, \text{L mol}^{-1} \, \text{s}^{-1}$, the question arises as to whether or not the data are consistent with an outer sphere mechanism. Since the calculated k_{12}, assuming $f_{12} = 1$, is $k_{12} = [(8.2 \times 10^2)(40)(2.6 \times 10^5)]^{1/2} = 9.2 \times 10^4 \, \text{L mol}^{-1} \, \text{s}^{-1}$, which is close to the experimental value of $1.5 \times 10^4 \, \text{L mol}^{-1} \, \text{s}^{-1}$, the mechanism is most likely to be an outer sphere one.

Note that as K_{12} increases, the rate increases. Outer sphere electron transfer rates decrease exponentially with the distance of separation of the two metal centers. For example, cytochrome c is an electron-transfer metalloprotein that contains a heme iron in either a Fe(II) or a Fe(III) state. All evidence indicates that electron transfer between two Fe centers is a long-range outer sphere process in which the electron tunnels through the protein. A number of model systems involving electron transfer from cytochrome c and small molecules such as $[Ru(H_2O)_6]^{2+}$ have been found to be consistent with an outer sphere electron transfer mechanism. Since the tunneling distance is much greater than for simple molecules, the rate constants are correspondingly much smaller.

BIBLIOGRAPHY

1. Maton, A.; Hopkins, J.; McLaughlin, C. W.; Johnson, S.; Warner, M. Q.; LaHart, D.; Wright, J. D. *Human Biology and Health;* Prentice Hall: Englewood Cliffs, New Jersey, 1993.

2. Pauling, L. *The Nature of the Chemical Bond,* 3rd ed.; Cornell University Press: Ithaca, NY, 1960; pp 145−181.

3. Crabtree, R. H. *The Organometallic Chemistry of the Transition Metals,* 5th ed.; John Wiley & Sons: New York, 2009.

4. Kauffman, G. B. *Classics in Coordination Chemistry, Part 1;* Dover Publications: New York, 1968.

5. Schläfer, H. L.; Gliemann, G. *Basic Principles of Ligand Field Theory;* Wiley Interscience: New York, 1969.

6. Cotton, F. A.; Wilkinson, G.; Murillo, C. A.; Bochmann, M. *Advanced Inorganic Chemistry,* 6th ed.; Wiley & Sons: New York, 1999.

7. Wells, A. F. *Structural Inorganic Chemistry,* 5th ed.; Oxford University Press: Oxford, UK, 1984.

8. Miessler, G.; Fischer, P. J.; Tarr, D. A. *Inorganic Chemistry,* 5th ed.; Pearson: New York, 2013.

9. Connelly, N. G.; Damhus, T.; Hartshorn, R. M.; Hutton, A. T.; Renner, T. *Nomenclature of Inorganic Chemistry: Recommendations 2005;* Royal Society of Chemistry: Cambridge, UK, 2005.

10. Sanderson, A. A Permanent Magnet Gouy Balance. *Phys. Educ.* **1968,** *3* (5), 272−273.

11. Brucacher, L.; Stafford, F. Magnetic Susceptibility. *J. Chem. Educ.* **1962,** *39,* 574.

12. Basalo, F.; Johnson, R. C. *Coordination Chemistry: The Chemistry of Metal Complexes;* W.A. Benjamin, Inc.: New York, 1964; pp 40−44.

13. Rodgers, G. E. *Introduction to Coordination, Solid State, and Descriptive Inorganic Chemistry;* McGraw -Hill: New York, 1994.

14. Murrel, J. N.; Kettle, S. F. A.; Tedder, J. M. *The Chemical Bond,* 2nd ed.; John Wiley & Sons: New York, 1985.

15. Pauling, L. Valence Bond Theory in Coordination Chemistry. *J. Chem. Educ.* **1962,** *39* (9), 461.

16. Lima-de-Faria, J.; Hellner, E.; Liebau, F.; Makovicky, E.; Parthé, E. Report of the International Union of Crystallography Commission on Crystallographic Nomenclature Subcommittee on the Nomenclature of Inorganic Structure Types. *Acta Crystallogr. Sect. A.* **1990,** *46,* 1−11.

17. Greenwood, N. N.; Earnshaw, A. *Chemistry of the Elements,* 2nd ed.; Butterworth-Heinemann: Oxford, UK, 1997.

18. Petrucci, R. H. *General Chemistry Principles and Modern Applications,* 9th ed.; Pearson Prentice Hall: Upper Saddle River, NJ, 2002.

19. Sherman, A.; Sherman, S. J.; Russikoff, L. *Basic Concepts of Chemistry,* 5th ed.; Houghton Mifflin Company: Boston, MA, 1992.

20. Johnson, R. C. A Simple Approach to Crystal Theory. *J. Chem. Educ.* **1965,** *42* (3), 147−148.

21. Pearson, R. G. Crystal Field Theory and Substitution Reactions of Metal Ions. *J. Chem. Educ.* **1961,** *38* (4), 164−173.

22. Silbeberg, M. *Chemistry: The Molecular Nature of Matter and Change,* 4th ed.; McGraw Hill Company: New York, 2006; pp 1028−1034.

23. Eisberg, R.; Resnick, R. *Quantum Physics of Atoms, Molecules, Solids, Nuclei, and Particles,* 2nd ed.; John Wiley & Sons: New York, 1985.

24. Cotton, F. A. *Chemical Applications of Group Theory;* Wiley-Interscience: New York, 1971.

25. Kealy, T. J.; Pauson, P. L. A New Type of Organo-iron Compound. *Nature* **1951,** *168,* 1039.

26. Lein, M.; Frunzke, J.; Timoshkin, A.; Frenking, G. Iron Bispentazole Fe(η^5-N_5)$_2$, a Theoretically Predicted High-energy Compound: Structure, Bonding Analysis, Metal-ligand Bond Strength and a Comparison with the Isoelectronic Ferrocene. *Chem. Eur. J.* **2001,** *7,* 4136−4155.

27. Pearson, R. G. The Nature of Substitution Reactions in Inorganic Chemistry. *J. Phys. Chem.* **1959,** *63* (3), 321−326.

28. Atwood, J. D. *Inorganic and Organometallic Reaction Mechanisms,* 2nd ed.; Wiley-VCH: Weinheim, DE, 1997.

29. Helm, L.; Merbach, A. E. Inorganic and Bioinorganic Solvent Exchange Mechanisms. *Chem. Rev.* **2005,** *105* (6), 1923−1960.

30. Atkins, P.; de Paula, J. *Physical Chemistry for the Life Sciences,* 2nd ed.; W.H. Freeman: London, UK, 2011.

31. Higginson, W. C. E.; Rosseinsky, D. R.; Stead, J. B.; Sykes, A. G. Mechanisms of Some Oxidation-reductions Between Metal Cations in Aqueous Solution. *Discuss. Faraday Soc.* **1960,** *29,* 49−59.

32. Atkins, P.; de Paula, J. *Physical Chemistry,* 10th ed.; Oxford University Press: Oxford, UK, 2014.

33. Kleinberg, R. *Mechanisms of Inorganic Reactions;* American Chemical Society: Washington, DC, 1965.

34. Quagliano, J. V.; Schubert, L. The Trans Effect in Complex Inorganic Compounds. *Chem. Rev.* **1952,** *50* (2), 201−260.

35. Coe, B. J.; Glenwright, S. J. Trans-effect in Octahedral Transition Metal Complexes. *Coord. Chem. Rev.* **2000,** *203,* 5−80.

36. Asperger, S. *Chemical Kinetics and Inorganic Reaction Mechanisms,* 2nd ed.; Springer Science & Business Media: New York, 2003.

37. Marcus, R. A. Electron Transfer Reactions in Chemistry Theory and Experiment. *J. Electroanal. Chem.* **1997,** *438,* 251−259.

38. Hush, N. S. Adiabatic Theory of Outer Sphere Electron-transfer Reactions in Solution. *Trans. Faraday Soc.* **1961,** *57,* 557−580.

39. Franck, J. Elementary Processes of Photochemical Reactions. *Trans. Faraday Soc.* **1926,** *21,* 536−542.

40. Condon, E. A Theory of Intensity Distribution in Band Systems. *Phys. Rev.* **1926,** *28,* 1182−1201.

41. Coolidge, A. S.; James, H. M.; Present, R. D. A Study of the Franck-Codon Principle. *J. Chem. Phys.* **1936,** *4,* 193−211.

42. Bernath, P. F. *Spectra of Atoms and Molecules,* 2nd ed.; Oxford University Press: Oxford UK, 2005.

Part 3

Advanced Topics-2: Electronic Spectra, Clusters & Isolobal Fragments

Coordination Chemistry: Electronic Spectra

1. INTRODUCTION: WHY DO WE NEED TO LEARN ELECTRONIC SPECTRA?

To understand reaction mechanisms, we need to know whether there is any formation of an intermediate that can be identified, even if it has only a very short period of detection time before being consumed to form the observed product. This can be accomplished through molecular spectroscopy, using a sensitive technique which collects vibrational and electronic spectral data of reaction intermediates. Herzberg's book entitled *Molecular Spectra and Molecular Structure. III. Electronic Spectra and Electronic Structure of Polyatomic Molecules* (Van Nostrand, Princeton, NJ, 1966) elegantly describes the spectroscopic detection system for reaction intermediates. It was the main source of information for many years, until the Chemistry WebBook database came into existence in the 1980s to 1990s, covering matrix isolation measurements, gas-phase electronic spectra, flash photolysis, photoelectron spectroscopy, etc. (http://webbook.nist.gov/chemistry/polyatom/). The excited electronic states helped to identify the molecular symmetry, the point group, and, eventually, they allowed us to determine the molecular structure of the stable product. This chapter takes you from the d-orbital splitting to the selection rules for a number of possible transitions involved in molecular spectroscopy.

2. ELECTRONIC SPECTRA

2.1 Selection rules

Selection rules govern what transitions can occur (are *allowed*) and what transitions cannot occur (are *forbidden*). It is also based on the symmetries of the orbitals.

Advanced Inorganic Chemistry. http://dx.doi.org/10.1016/B978-0-12-801982-5.00008-4

In atomic spectroscopy, the selection rules are: $\Delta S = 0$; $\Delta L = 0, \pm 1$; $\Delta l = \pm 1$; $\Delta J = 0, \pm 1$ (but not $J = 0 \Leftrightarrow J = 0$). For the doublet D line of Na, both transitions, shown below, are allowed and, therefore, they are very intense.

$^2P_{3/2} \rightarrow {}^2S_{1/2}$ at 589.16 nm, $\Delta S = 0$, $\Delta L = -1$, $\Delta l = -1$, $\Delta J = -1$.

$^2P_{1/2} \rightarrow {}^2S_{1/2}$ at 589.76 nm, $\Delta S = 0$, $\Delta L = -1$, $\Delta l = -1$, $\Delta J = 0$.

In molecular spectroscopy, the important selection rules are:

1. **Spin selection rule.** $\Delta S = 0$

$$\text{spin allowed} \quad \begin{cases} \text{singlet} \Leftrightarrow \text{singlet} \\ \text{triplet} \Leftrightarrow \text{triplet} \end{cases}$$

$$\text{spin forbidden} \quad \text{singlet} \Leftrightarrow \text{triplet}$$

2. **La Porte selection rule.** If molecule has a center of inversion, $\Delta l = \pm 1$

$$\text{La Porte allowed} \quad \begin{cases} d \Leftrightarrow p \\ d \Leftrightarrow f \end{cases}$$

$$\text{La Port forbidden} \quad d \Leftrightarrow d$$

The visible spectra of transition metal complexes are due to the transition from a low-energy d orbital to a higher energy one. Since octahedral transition metal complexes have centers of inversion, they should all be colorless.

3. **Vibronic coupling.**

An electronic transition always can also give rise to a change in vibrational state. Therefore, one should consider $\psi_{el}\psi_{vib}$ instead of just ψ_{el}. In a transition from state (1) $\rightarrow$ state (2) while $\psi_{el}(1)$ and $\psi_{el}(2)$ might not be symmetry compatible, $\psi_{el}\psi_{vib}(1)$ and $\psi_{el}\psi_{vib}(2)$ could be compatible.

During an asymmetric stretching vibration, the center of inversion could be momentarily destroyed and the La Porte rule is not applicable. This is a qualitative explanation for the observed behavior. The d $\Leftrightarrow$ d transitions are said to be *vibronically allowed*. Because of this, the peaks in the spectrum of octahedral complexes are very broad compared with allowed transitions that occur in tetrahedral complexes.

4. **Molar absorptivity** ($\in$).

In octahedral complexes, molar absorptivity ($\in$) ranges from a low of 0.01 for high-spin d^5 (Mn^{2+}, Fe^{3+}) (both La Porte and spin forbidden) to a high of 25–30 for low-spin Co(III) complexes. In solution, "forbidden" transitions can occur, but at very weak molar absorptivity.

In tetrahedral complexes, ϵ's are much higher, ranging from 50 to 200. These are all small compared with the ϵ's $\approx 10^4$ for the allowed charge transfer transitions.

2.2 Spectra of octahedral (O_h) and tetrahedral (T_d) complexes

1. The effect of the crystal field is to remove the degeneracy of the term symbol atomic states of the metal (Table 8.1).

Table 8.1 Atomic States and Corresponding Term Symbols

L	Atomic State	$2L+1$	States in an O_h or a T_d Field
0	S	1	A_1
1	P	3	T_1
2	D	5	$E + T_2$
3	F	7	$T_1 + T_2 + A_2$
4	G	9	$A_1 + E + T_1 + T_2$
5	H	11	$E + 2T_1 + T_2$
6	I	13	$A_1 + A_2 + E + T_1 + 2T_2$

The states in an O_h or T_d field are the standard group theory notations. Recall that an "A" state is nondegenerate, an "E" state is doubly degenerate, and a "T" state is triply degenerate. Since the O_h point group has a center of inversion, each state can be classified as *gerade* (*g*) if it does not change sign on inversion or *ungerade* (*u*) if it does. All d orbitals have "*g*" symmetry, while p orbitals have "*u*" symmetry. These are written as right-hand subscripts. Example: In an O_h field a D state is transformed into an E_g and a T_{2g}.

2. Ground state term symbols for transition metals are:

$$d^1 = d^9 = {}^2D \quad d^2 = d^8 = {}^3F \quad d^5 = {}^6S$$

$$d^4 = d^6 = {}^5D \quad d^3 = d^7 = {}^4F \quad d^{10} = {}^1S$$

Note ground states are either S, D or F.

3. Energy differences between states can be explained as follows.

   a. States arising from the same atomic term symbol states will differ in energy in units of Dq.

   b. States arising from different atomic (term symbol) states will differ in energy in units of Dq + the difference in energy of the atomic states.

   c. Energies between states can be expressed using two parameters, **B** and **C**, called Racah parameters. Thus, if the atomic states have the same spin, only **B** is involved, and if the states have different spins, then both **B** and **C** are involved.

   d. Dq, **B,** and **C** are obtained experimentally by analyzing the spectra of the complexes. Any transition involving Dq will be broad due to vibrations slightly changing the value of Dq.

2.3 Orgel diagrams

High-spin octahedral and tetrahedral d^1, d^4, d^6, and d^9 complexes are all arising from D states and they can be shown in an Orgel diagram (Fig. 8.1).

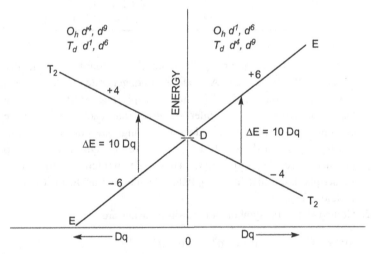

■ **FIGURE 8.1** Orgel diagram with a single transition energy of 10 Dq.

In all cases, only a single transition of energy 10 Dq can be expected.

1) O_h	d^1	$^2E_g(D)$	$^2T_{2g}(D)$	d^9	$^2T_{2g}(D)$	$^2E_g(D)$
	d^6	$^5E_g(D)$	$^5T_{2g}(D)$	d^4	$^5T_{2g}(D)$	$^5E_g(D)$
2) T_d	d^1	$^2T_2(D)$	$^2E(D)$	d^9	$^2E(D)$	$^2T_2(D)$
	d^6	$^5T_{2g}(D)$	$^5E_g(D)$	d^4	$^5E(D)$	$^5T_2(D)$

2.3.1 Simple one-electron approach

1. d^1 System.

O_h

Three ways

∴ T_{2g}

E = −4 Dq

Two Ways

∴ E_g

E = +6 Dq

T_d

Two ways

∴ E_g

E = −6 Dq

Three ways

∴ T_{2g}

E = +4 Dq

2. High-spin d^6 is exactly the same as d^1 if half-filled orbital arrangements do not change.

3. d^4 System.

O_h

Two ways

∴ E_g

E = −6 Dq

Three ways

∴ T_{2g}

E = +4 Dq

T_d

Three ways

∴ T_{2g}

E = −4 Dq

Two ways

∴ E_g

E = +6 Dq

4. The d^9 transitions are the same as d^4 since each orbital is half-filled.

2.3.2 Effect of distortion

1. O_h complexes : d^1 and high spin d^6 have slight distortion.
   d^4 (high spin) and d^9 have high distortion.

2. General effect of distortion is to D_{4h} symmetry (Fig. 8.2).

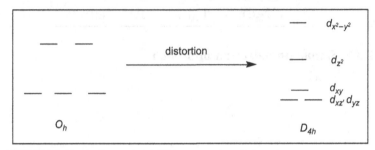

■ **FIGURE 8.2** Distortion from octahedral (O_h) into square-planar (D_{4h}) geometry.

3. Slight distortion gives rise to a broadening of the transition to show only one peak.
4. Large distortion depends on the extent of the distortion. The "completely" distorted square planar complex would have three transitions as in the d^4 system, shown in Fig. 8.3.

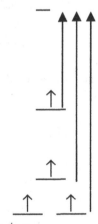

■ **FIGURE 8.3** Possible transitions in d^4 system.

2.3.3 Systems arising from F states

1. High-spin O_h and T_d d^1, d^3, d^7, and d^8 complexes.
   Recall that the effect of the ligand field is to split the F state into A_2, T_1, and T_2. Therefore, transitions among these states and any other low lying energy states must be considered.

2. In the d^2 system, from atomic spectroscopy, the states are: $^3F < {}^1D < {}^3P < {}^1G < {}^1S$. Therefore, the effect of the crystal field will be:

$$^3P \rightarrow {}^3T_1$$
$$^3F \rightarrow {}^3A_2 + {}^3T_1 + {}^3T_2$$

Transitions among these are all spin allowed and they must be considered.

3. Note that there are two $^3T_1$ states. States of the same symmetry cannot cross (noncrossing rule). Therefore, the two states will interact, one increasing in energy, the other decreasing in energy. The closer the energy of the two states, the stronger will be the interaction. This causes the two states to be nonlinear in Dq.

4. If the two T_2 states did not interact, then the Orgel diagram shown in Fig. 8.4 with "B" as a Racah parameter, will be expected.

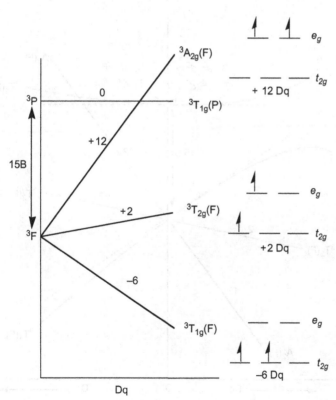

■ **FIGURE 8.4** Orgel diagram for the two noninteracting T_2 states.

Accordingly, one would expect three transitions:

$$^3T_{2g}(F) \leftarrow {}^3T_{1g}(F) \quad \Delta E = 8\,Dq$$
$$^3T_{1g}(P) \leftarrow {}^3T_{1g}(F) \quad \Delta E = 15B + 6\,Dq$$
$$_3A_{2g}(F) \leftarrow {}^3T_{1g}(F) \quad \Delta E = 18\,Dq$$

5. The $^3A_{2g}(F)$ and $^3T_{2g}(F)$ are linear with Dq and are the same as above. The two $^3T_{1g}$ states have energies that are roots (solutions) of the determinant.

$$\begin{vmatrix} -6\,Dq - E & 4\,Dq \\ 4\,Dq & 15B - E \end{vmatrix} = 0$$

This causes E to be nonlinear in Dq. The Orgel diagram for O_h and T_d complexes with d^2, d^3, d^7, and d^8 configurations is shown in Fig. 8.5.

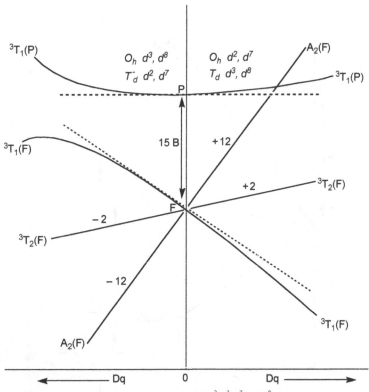

■ **FIGURE 8.5** Orgel diagram for O_h and T_d complexes of d^2, d^3, d^7, and d^8 configurations.

The transitions are:

$$O_h\left(d^2, d^7\right) \begin{cases} T_{2g}(F) \leftarrow T_{1g}(F) \\ T_{1g}(P) \leftarrow T_{1g}(F) \\ A_{2g}(F) \leftarrow T_{1g}(F) \end{cases} \quad T_d\left(d^3, d^8\right) \begin{cases} T_2(F) \leftarrow T_1(F) \\ T_1(P) \leftarrow T_1(F) \\ A_2(F) \leftarrow T_1(F) \end{cases}$$

$$O_h\left(d^3, d^8\right) \begin{cases} T_{2g}(F) \leftarrow A_{2g}(F) \\ T_{1g}(F) \leftarrow A_{2g}(F) \\ T_{1g}(P) \leftarrow A_{2g}(F) \end{cases} \quad T_d\left(d^2, d^7\right) \begin{cases} T_2(F) \leftarrow A_2(F) \\ T_1(F) \leftarrow A_2(F) \\ T_2(P) \leftarrow A_2(F) \end{cases}$$

None of the transitions originating from the $T_1(F)$ ground state (O_h d^2, d^7 and T_d d^3, d^8) can be written in terms of Dq, since the state is nonlinear in Dq. The lowest energy transition in the O_h d^3, d^8 and T_d d^2, d^7 is exactly equal to 10 Dq.

Approximate analysis of the spectra of O_h d^3, d^8 and T_d d^2, d^7 can be made accordingly.

1. In cases where Dq is small (weak field) the energies can be approximated by:

$$A_2(F) = -12\,Dq\,(\text{exact})$$
$$T_2(F) = -2\,Dq\,(\text{exact})$$
$$\left.\begin{array}{l} T_1(F) = +6\,Dq - x \\ T_1(P) = 15B + x \end{array}\right\} \text{Assuming that the splitting is symmetric.}$$

2. Transition energies (v) are wave numbers (cm^{-1}).

$$v_1 = 10\,Dq \qquad\qquad T_2(F) \leftarrow A_2(F)$$
$$v_2 = 18\,Dq - x \qquad\quad T_1(F) \leftarrow A_2(F)$$
$$v_3 = 12\,Dq + 15B + x \quad T_1(P) \leftarrow A_2(F)$$
$$\therefore Dq = \frac{v_1}{10}; v_2 + v_3 = 30\,Dq + 15B = 3v_1 + 15B$$
$$\therefore v_2 + v_3 - 3v_1 = 15B$$

3. Some of the examples include the spectrum of $Ni(H_2O)_6{}^{2+}(d^8)$ showing three peaks at:

$$v_1 = 8600\ cm^{-1} = 10\,Dq$$
$$v_2 = 13,500\ cm^{-1} = 18\,Dq - x$$
$$v_3 = 25,300\ cm^{-1} = 12\,Dq + 15B + x$$
$$\therefore Dq = \frac{8600\ cm^{-1}}{10} = 860\ cm^{-1}$$
$$15B = 25,300\ cm^{-1} + 13,500\ cm^{-1} - 3\left(8600\ cm^{-1}\right) = 13,000\ cm^{-1}$$
$$B = \frac{13,000\ cm^{-1}}{15} = 867\ cm^{-1}$$

Similarly, the spectrum of $Ni(NH_3)_6^{2+}$ is showing three peaks at:

$$\nu_1 = 10,700 \text{ cm}^{-1}$$
$$\nu_2 = 17,500 \text{ cm}^{-1}$$
$$\nu_3 = 28,200 \text{ cm}^{-1}$$

The calculation shows $Dq = 1070 \text{ cm}^{-1}$; $B = 907 \text{ cm}^{-1}$.

4. Note that Dq for NH_3 is greater than that of H_2O as shown in the Fajan–Tsuchida series and obtained from measurements of the lowest energy transitions for the Ni^{2+} octahedral complexes of the different ligands.

5. The Recah parameter, B, differs slightly in going from H_2O to NH_3. Since it depends on the difference in energy between the ^3F and the ^3P state, it should be independent of the ligand, and is the same as that obtained from atomic spectroscopy of the metal. However, the Recah parameter, B, will differ slightly for a particular metal.

The change in B with the ligand is known as the *nephelauxetic effect* (nephelauxetic = cloud expansion). This is because the interaction between the metal and the ligands is not purely ionic (as assumed in crystal field theory). The metal forms covalent bonds [molecular orbitals (MOs)] with the ligands. The resulting MOs are more extended out in space than are the metal d-orbitals. This reduces the apparent energy difference between the ^3F and ^3P states. This led to an extension of the electrostatic (Crystal Field Theory) to include covalent interactions (Ligand Field Theory). The Ligand Field Theory was very similar to Molecular Orbital Theory, and has been replaced by the latter.

6. High-spin d^5 systems.

High-spin d^5 systems have a ^6S ground state configuration that transforms in an octahedral or tetrahedral field to a $^6A_1$ state, whose energy is independent of Dq. The atomic states for the system are: $^6S < {}^4G < {}^4P < {}^4D < {}^4F$. A partial Orgel diagram is shown in Fig. 8.6. All energies are relative to the $^6A_1$ state.

7. The spectrum should consist of a number of weak spin forbidden transitions, with the most intense being the $^4E_g, {}^4A_{1g} \leftarrow {}^6A_1$ transition. Since the energies of the ^4E_g, $^4A_{1g}$, and $^6A_1$ states do not change with Dq, this peak is very sharp. All Mn^{2+} salts are very pale pink in color due to the $^4E_g, {}^4A_{1g} \leftarrow {}^6A_1$ transition at about 25,000 cm^{-1}. Due to the nephelauxetic effect, its position will vary slightly from complex to complex. The absorption spectrum of a single crystal of MnF_2 is shown in Fig. 8.7.

8. Low-spin d^6 systems.

Some of the most important and numerous transition metal complexes are the octahedral complexes of Co(III) ion. Except for

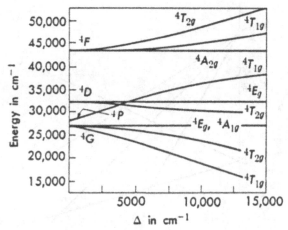

■ **FIGURE 8.6** A partial Orgel diagram for the Mn^{2+} ion.

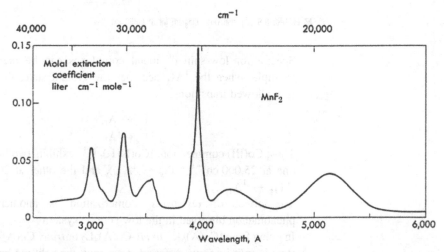

■ **FIGURE 8.7** The absorption spectrum of a single crystal of MnF_2 at 25°C.

complexes with very weak field ligands, such as CoF_6^{3-}, most are low-spin species.

The atomic states for the d^6 system are: $^5D < {}^3H < {}^3P < {}^3F < {}^3G < {}^1I < {}^3G < {}^1G$. A partial Orgel diagram is shown in Fig. 8.8. One can notice that as Dq increases, the states arising from the ^1H dropped rapidly in terms of energy.

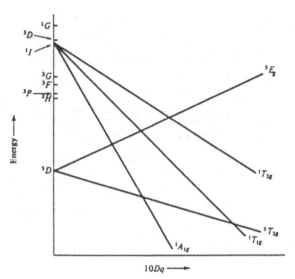

■ **FIGURE 8.8** A partial Orgel diagram for d^6 Co(III) ion.

Spectra for low-spin d^6 metal complexes can be predicted. For example, when the $^1A_{1g}$ becomes the ground state, there are two spin allowed transitions:

$$^1T_{1g} \leftarrow {}^1A_{1g}$$
$$^1T_{2g} \leftarrow {}^1A_{1g}$$

Thus, Co(III) complex ion, $[Co(NH_3)_6]^{3+}$, exhibits two broad peaks, one at 25,000 cm^{-1} ($^1T_{2g} \leftarrow {}^1A_{1g}$), and the other at 20,000 cm^{-1} ($^1T_{1g} \leftarrow {}^1A_{1g}$).

The effect of *cis–trans* isomerization is another common phenomenon observed in the low-spin d^6 system.

In going from $[Co(A)_6]^{3+}$ to *cis*-Co(A)$_4$L$_2$ or *trans*-Co(A)$_4$L$_2$, the T_{1g} state is split into two different states with the splitting being greater for the *trans* isomer (Fig. 8.9).

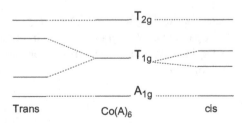

■ **FIGURE 8.9** Effect of *cis–trans* isomerization in d^6 Co(III) complex.

The splitting just introduces a broadening in the *cis* isomer. However, in the *trans* isomer, the splitting is large enough so that three peaks are observed. This can be used as a proof of structure for the *trans* isomer.

3. TANABE—SUGANO DIAGRAMS

Coordination chemistry utilizes Tanabe—Sugano diagrams to predict absorptions in the ultraviolet, visible, and infrared spectrum of compounds. The Tanabe—Sugano diagram analysis of any coordination compound is correlated with the observed spectroscopic data, as they are qualitatively useful to approximate the 10 Dq value of the ligand field splitting energy.

1. Plots of energy versus Dq for various d^n systems should help to interpret the observed spectra. A modified Tanabe—Sugano diagram for a d^6 [Co(III)] metal in an octahedral field is shown in Fig. 8.10.
2. It is similar to an Orgel diagram except that (1) E and Dq are plotted in terms of the Racah parameter, B, and (2) the ground state is plotted as the horizontal (x) axis, whether or not it changes with Dq. Those changes are reflected in the energies of the excited states where energies are all relative to the ground state.
3. If the ground state energy changes as Dq changes, it is absorbed in the higher energy states.
4. Example of the use of Tanabe—Sugano diagrams.
   Diamagnetic $Co(en)_3^{3+}$ exhibits two peaks, $v_1 = 21,400$ cm^{-1} and $v_2 = 29,500$ cm^{-1}, corresponding to the $^1T_{1g} \leftarrow {}^1A_{1g}$ and the $^1T_{2g} \leftarrow {}^1A_{1g}$. transitions.
5. The general steps for using the Tanabe—Sugano diagrams are as follows:
   a. Determine the ratio of the energies of the two transitions.
      $v_2/v_1 = 29,500$ cm^{-1}/21,400 cm$^{-1} = 1.38$.
   b. Go to the diagram and find the value of Dq/B that gives this ratio.
   c. Read the energies of the $^1T_{1g}$ and $^1T_{2g}$ states at that value of Dq/B. The two energies will be in terms of E/B. Since E is the transition energy (29,500 cm^{-1} or 21,400 cm^{-1}), the value of B can be obtained. Once B is known, Dq can be determined.
   d. The scale on the diagram shown in Fig. 8.10 is so large that too much error would be introduced in using this method. In actual practice, computer programs are used to obtain the exact values of Dq/B and E/B. A careful analysis of the $[Co(NH_3)_6]^{3+}$ gives Dq = 2287 cm^{-1} and B = 615 cm^{-1}.

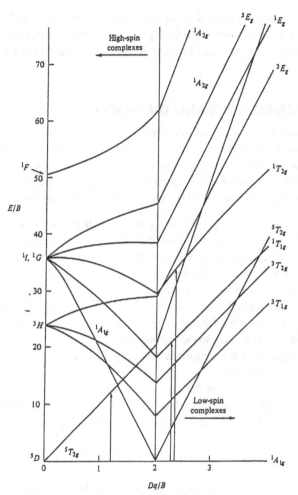

■ **FIGURE 8.10** A modified Tanabe—Sugano diagram for a d^6 [Co(III)] complex.

4. CHARGE TRANSFER SPECTRA

We have been considering transitions between states arising from the d orbitals in the various ligand fields. It is also possible to have transitions of ligand-based electrons to the metal and vice versa. Such transitions are called *charge transfer* transitions. These transitions are responsible for the colors of some d^0 (purple MnO_4^-) and d^{10} (brick red HgI_2, orange red BiI_3).

Charge transfers are very intense transitions with molar extinction coefficients of $\sim 50,000$, compared with ~ 200 or less for d–d transitions. In general, there are two types of transitions that can give rise to charge transfer bands.

1. An electron in a ligand-based σ bonding orbital can make a transition to a vacant, metal-based orbital (in O_h complexes the e_g^* or in T_d complexes the t_2). This transfers charge from the ligand to the metal and is called a *charge transfer to metal* or *ligand to metal charge transfer* (LMCT). Note that in such a transition, the metal is "reduced" and the ligand is "oxidized".

   a. The charge transfer spectra of the heavy metal hexahalide complexes have been studied in detail by Jorgensen, who found three main regions in their absorption spectrum. A group of narrow bands are observed between 15,000 and 30,000 cm^{-1} that increase in energy in the sequence $I < Br < Cl$ in the nd^3, nd^4, and nd^5, but not in low-spin nd^6. All hexa-halogen anions of the 4d and 5d elements exhibit broad, strong adsorptions in the region 25,000–45,000 cm^{-1}. In certain cases, another very strong band above 44,000 cm^{-1} is also present.

2. These features can be understood using the simplified MO diagram in Fig. 8.11 showing the possible LMCT transitions for a single halide ligand.

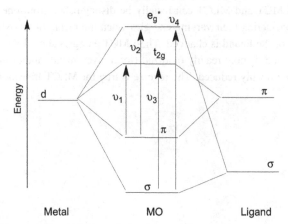

■ **FIGURE 8.11** Molecular orbital diagram for ligand to metal charge transfer in complex.

Recall that the halides have filled p_π electrons so that the lower energy σ and π MOs are filled in ligand-centered orbitals, while the e_g and t_{2g} are metal centered. The possible transitions are:

ν_1	$\pi(L)$	$t_{2g}(M)$	ν_2	$\pi(L)$	$e_g(M)$
ν_3	$\sigma(L)$	$t_{2g}(M)$	ν_4	$\sigma(L)$	$e_g(M)$

The lowest energy transitions can be assigned to ν_1 and they should not be present in the low-spin nd^6 complexes, since the t_{2g} orbitals are filled. The energies should also increase as the ionization potential of the halogen increases. The lowest energy set of transitions for the nd^6 complexes is assigned to ν_2.

The highest energy set is assigned to ν_4 and is present in the nd^6 complexes, which would rule out ν_3. The actual MO diagram for all six ligands is more complex than the one shown in Fig. 8.11 where several peaks are possible, despite a single transition being shown in this figure.

3. The other type of charge transfer transition would be that involving an electron in a metal-based orbital making a transition to a low lying, vacant ligand-based orbital. Such a transition is called a *charge transfer to ligand* or *metal to ligand charge transfer* (MLCT). The MLCT transitions would be expected for ligands having empty π^* antibonding orbitals, such as CN^-, pyridine, 2,2'-bipyridine, *o*-phenanthroline, etc., and metals that are easily oxidized.

4. The LMCT and MLCT can usually be distinguished from one another by considering their varying energies when the metal, or its oxidation state, or the ligand is changed. The LMCT energies should decrease as the ligand is more readily oxidized (for a given metal) and/or the metal is more readily reduced. The reverse is true for MLCT transitions.

Cluster Chemistry and Isolobal Fragments

1. INTRODUCTION: ROLE OF CLUSTER CHEMISTRY IN NATURE

Robust cluster formation is Nature's creation by incorporating main group elements and transition metals, alone as in "naked" clusters or in "organometallics" that are stabilized by ligands, or even consisting of only one kind of molecule, such as water clusters. Thus, one can see atomic clusters, molecular clusters, transition metal halide and organometallic clusters, boron hydrides, protein-bound iron-sulfur clusters ([Fe-S]), Zintl clusters, metalloid clusters, fullerenes, and many more. The fascinating aspects of these clusters are those where nonmagnetic elements can become magnetic, semiconducting materials can exhibit metallic behavior, metallic systems may turn into semiconducting, the color of particles could change with size, noble metals could be reactive, and brittle materials are made malleable, and many more. These unusual reactivity patterns are due to unusual cluster geometry in which electrons exhibit an energy gap between the highest occupied molecular orbital (HOMO) and the lowest unoccupied molecular orbital (LUMO). The magnitude of the HOMO-LUMO gap depended upon the size and cluster geometry leading to stability and reactivity of the clusters. A systematic study of the structure and properties of clusters comprising various elements has encompassed many areas of chemistry, biochemistry, and also physics, particularly of atomic, molecular, nuclear, and condensed-matter physics.

Advanced Inorganic Chemistry. http://dx.doi.org/10.1016/B978-0-12-801982-5.00009-6

2. CLUSTERS OF BORANES, CARBORANES, AND THEIR METAL COMPLEXES

2.1 Terminology used in polyhedral boron clusters

Boranes are mixed hydrides of boron in electron-deficient cages. Let us look at the geometries and the corresponding bonding in ethane (C_2H_6) and diborane (B_2H_6) (Fig. 9.1).

The B_2H_6 molecule is held together by two three-centered two-electron (3c-2e) B-H-B bridge bonds and four two-centered two-electron (2c-2e) normal B-H sigma bonds. Thus, there are two types of hydrogens: four terminal and two bridging hydrogens.

1. All boranes are characterized by multicentered bonds, as well as terminal and bridging hydrogens. Fig. 9.2 shows the structures of some of the simpler boranes.
2. The boranes in Fig. 9.2, other than diborane, have triangular faces and can be considered as clusters of fused deltahedra. A boron atom, along with at least one terminal hydrogen, forms the vertices of a cage, which may also have bridging hydrogens. For example, B_5H_9 (pentaborane) has five boron atoms at the vertices of a square pyramidal cage. Each boron is bonded to an *exo*-polyhedral hydrogen (the terminal hydrogens); in addition, there are four hydrogens bridging adjacent borons on the rectangular open face of the cluster.
3. The bridging hydrogens are not as strongly bonded as are the terminal ones and can be removed by strong bases such as butyllithium (BuLi) or sodium hydride (NaH). The anion $[B_3H_8]^-$ in Fig. 9.2 is prepared by removing one of the bridging hydrogens in B_3H_9.
4. Some of the boranes, such as $[B_6H_6]^{2-}$ and $[B_8H_8]^{2-}$ dianions have completely closed (*closo*) polyhedral cages.

C_2H_6

14 valence electrons

14 valence orbitals

electron precise

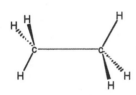

Seven *2c-2e* bonds

B_2H_6

12 valence electrons

14 valence orbitals

electron deficient

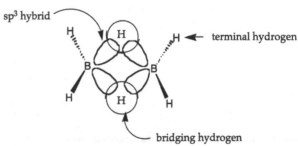

■ **FIGURE 9.1** Bonding and structures of ethane and diborane.

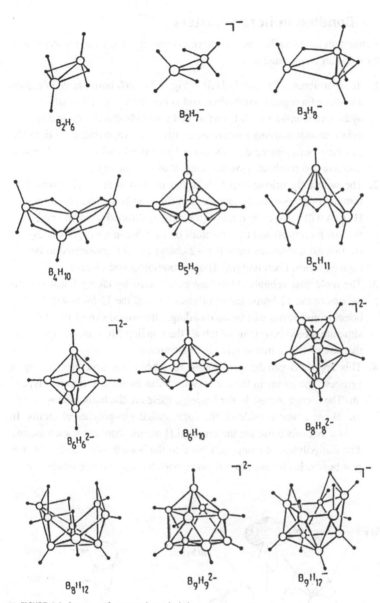

■ **FIGURE 9.2** Structures of common boron hydrides.

2.2 **Bonding in boron clusters**

Bonding in the *closo*-boranes has been extensively studied and it serves as a basis for all other structures.

1. In the structure of $[closo\text{-}B_4H_4]^{2-}$ (Fig. 9.3), each boron atom occupies a vertex of a regular tetrahedron and is bonded to one terminal hydrogen. Consider each boron as being sp hybridized with one sp hybrid orbital pointing toward the center of the cage (radially oriented) and the other pointing directly toward a terminal hydrogen. Each boron also has two p orbitals oriented tangentially to the cage.

2. The outwardly oriented sp hybrid can overlap with an H s orbital to form a 2c-2e bond, using one electron from the boron and one from the H. This will leave three orbitals (the radially directed sp hybrid and the two p orbitals) and two electrons on each boron for cage bonding, the two extra electrons from the −2 charge are also considered to be cage electrons. Each bridging H also contributes one electron.

3. The molecular orbitals (MOs) are constructed by taking linear combinations of the 12 boron-based orbitals. Five of the 12 MOs will be bonding and seven will be antibonding. One would expect the stable structure would be one in which all the bonding MOs are occupied. In the case of B_4H_4, that would be 10 electrons.

4. This approach can be generalized to other *closo*-structures having n vertices. The atoms in these compounds can be divided into two types.
    a. The n cage atoms. In the boranes, these are the boron atoms.
    b. Those atoms outside of the cage, called *exo*-polyhedral atoms. In the boranes these are the terminal H atoms, there are n such atoms.

5. The *exo*-polyhedral atoms will bond to the borons using 2c-2e electron pair bonds. Each cage atom donates one electron and one orbital for

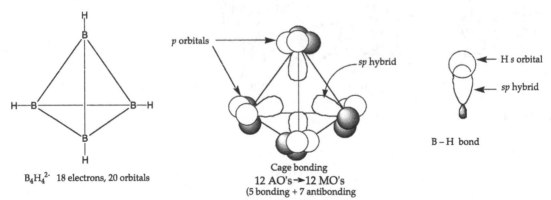

$B_4H_4^{2-}$ 18 electrons, 20 orbitals

Cage bonding
12 AO's→12 MO's
(5 bonding + 7 antibonding)

■ **FIGURE 9.3** Bonding molecular orbitals (MOs) in $[closo\text{-}B_4H_4]^{2-}$ cluster. *AO*, atomic orbital.

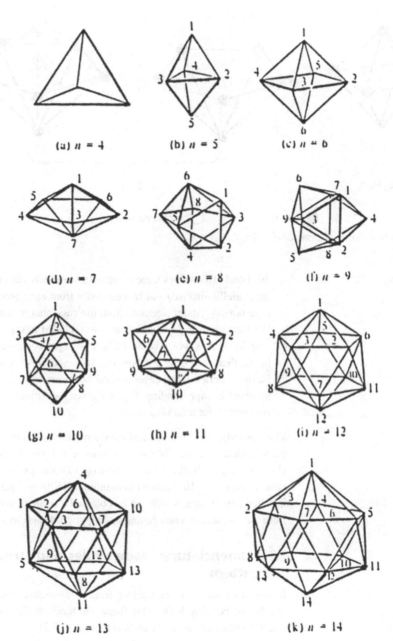

■ **FIGURE 9.4** Simple *closo*-boranes with the formula $[B_nH_n]^{2-}$.

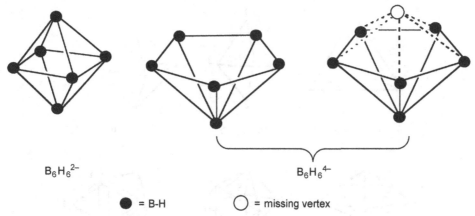

$B_6H_6^{2-}$

$B_6H_6^{4-}$

● = B-H   ○ = missing vertex

■ **FIGURE 9.5** *closo-* and *nido-*structures of anionic $[B_6H_6]^{n-}$ cluster ($n = 2$ or 4).

this bond. This leaves three orbitals (one radially directed and two tangentially directed) and two electrons from each boron vertex for cage bonding; extra electrons from the cage charge and from bridging H's (one electron from each) are also included in cage bonding.

6. The $3n$ orbitals will give $3n$ MOs. For cages composed of fused delta-hedra, there will be $n + 1$ bonding MOs and $2n-1$ antibonding MOs.

7. Stable n vertex *closo*-cages should be those with $n + 1$ electron pairs involved in cage bonding. Fig. 9.4 gives the structures and numbering convention for a number of cages.

The $[closo\text{-}B_6H_6]^{2-}$ has an octahedral structure that is consistent with its 26 valence electron count (14 cage electrons, $n + 1$ pairs). The addition of two electrons to give $B_6H_6^{4-}$ causes the cage to open, giving a pentagonal pyramidal cage with 16 electrons in bonding MOs ($n + 2$ pairs). The structure is called a *nido*-cage, which can be obtained if a single vertex was removed from a *closo* seven-vertex pentagonal bipyramid cage, as shown in Fig. 9.5.

2.3 Nomenclature: Wade's rules and structural pattern

Removal of one or more vertices from a *closo*-cage does not change the number of bonding MOs. This forms the basis of the polyhedral electron counting rules known as Wade's rules (Table 9.1).

Fig. 9.6 gives the progression of several important cages from *closo* (closed) to *nido* (nest-like) to *arachno* (cobweb-like). Further removal would lead to more open cages (*hypo*, *fisco*, *reticulo*). Usually adjacent vertices are removed when forming the more open structures.

Table 9.1 Structural Types and Nomenclature

	Closo	Nido	Arachno	Hypo	Fisco	Reticulo
Occupied vertices	n	n	n	n	n	n
Skeletal electron pairs	$n+1$	$n+2$	$n+3$	$n+4$	$n+5$	$n+6$

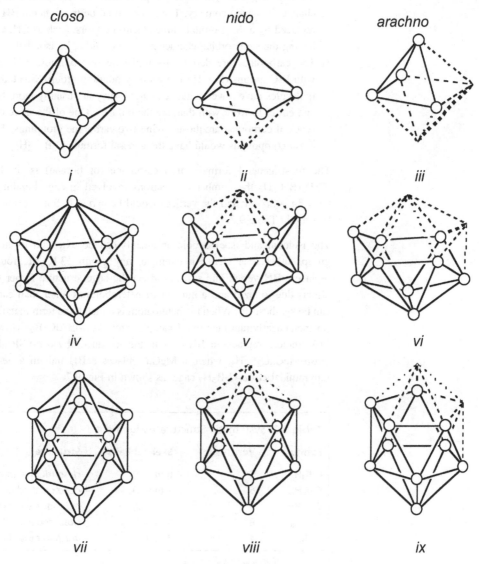

■ **FIGURE 9.6** Some common structures of *closo-*, *nido-*, and *arachno-*polyhedral cages.

The utility of the Wade's rules can be extended by noting the following:

1. The terminal H's could be replaced by any atom or group that forms a single bond (halogen, CH_3, NH_2, etc.) without changing any structural particulars.
2. The replacement of a BH vertex by any group that donates two electrons and has three suitably oriented orbitals (a radially directed orbital and two tangentially directed p-type orbitals) should not materially change the cage geometry. The same would be true when a BH^- is replaced by a three-orbital, three-electron donors, such as CH. Groups having the same orbital characteristics are said to be isolobal.
3. The carboranes are derived by replacing one or more BH^- vertices with R-C groups (R = H or any singly bonding group). Since these molecules have lower negative charges, they are usually more stable and easier to work with than are the boranes. One of the more common types of carborane are those having two carborane groupings. The *closo*-compounds would have the general formula $C_2B_{n-2}H_n$.

The most general formula of a carborane (or borane) is of the form $(CH)_a(BH)_bH_c$ the number of electrons involved in cage bonding would be $= 3a + 2b + c$ and the vertices would be $= a + b$. Some examples can be seen in Table 9.2.

The isolobal and isoelectronic arguments can be extended to atoms and groups other than CH. For example, any group 13 R—M group could replace a BH vertex and still not affect the cage structure (other than the effects due to size) and a number of heteroboranes other than carboranes can be synthesized. When the heteroatom is a metal, the term metallaborane (or metallacarborane) is used. Examples are: *closo*-EtAlC$_2$B$_9$H$_{11}$, where an EtAl moiety replaces a BH unit in the icosahedral *closo*-C$_2$B$_{10}$H$_{12}$ and *closo*-MeGaC$_2$B$_4$H$_6$, where a MeGa replaces a BH unit in a pentagonal bipyramidal *closo*-C$_2$B$_5$H$_7$ cage, as shown in Fig. 9.7.

Table 9.2 Systematic Structure Assignment

Formula	Vertices (*n*)	Skeletal Pairs	Structure
$C_2B_3H_5$	5	6 ($n + 1$)	*closo*-trigonal bipyramid
$C_2B_4H_6$	6	7 ($n + 1$)	*closo* octahedral
$C_2B_3H_7$	5	7 ($n + 2$)	*nido* square pyramid (*ii*)[a]
$C_2B_6H_8$	8	9 ($n + 1$)	*closo* dodecahedron
B_4H_{10}	4	7 ($n + 3$)	*arachno* butterfly (*iii*)[a]

[a]*Structure numbering is as in Fig. 9.6.*

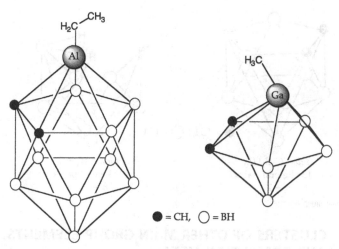

● = CH, ○ = BH

■ **FIGURE 9.7** Examples of group 13 metallacarboranes.

A group 14 element other than carbon atom could replace a BH^- unit and not affect the cage structure. In general, the underivatized heavier group 15 metals could substitute for a BH^- with their lone pairs being *exo*-polyhedral. For example, *closo*-$SnC_2B_4H_6$ has a pentagonal bipyramidal structure, related to *closo*-$C_2B_5H_7$.

Heteroboranes are those where the isolobal/isoelectronic fragments of other main group elements have been incorporated into the cluster as shown in Table 9.3.

Some examples can be seen in Fig. 9.8.

Table 9.3 Skeletal Electron Contributions of Main Group Cluster Units $(v+x-2)^a$

Group Number (v)	Element	Cluster Unit		
		M (x = 0)	MR (x = 1)	MR₂ or M-L (x = 2)
1	Li, Na	–	0	1
2	Be, Mg, Zn, Cd	0	1	2
13	B, Al, Ga, In, Tl	1	2	3
14	C, Si, Ge, Sn, Pb	0	3	4
15	N, P, As, Sb, Bi	3	4	–
16	O, S, Se, Te	4	5	–

^{a}v, No. valence electrons on M; x, No. electrons from ligands on M (R = one electron ligand or H; L = two-electron ligand).

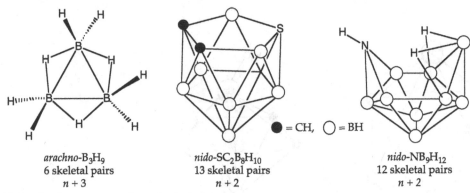

arachno-B_3H_9
6 skeletal pairs
$n+3$

nido-$SC_2B_8H_{10}$
13 skeletal pairs
$n+2$

● = CH, ○ = BH

nido-NB_9H_{12}
12 skeletal pairs
$n+2$

■ **FIGURE 9.8** Examples of open-cage borane clusters of main group elements.

3. CLUSTERS OF OTHER MAIN GROUP ELEMENTS AND TRANSITION METALS

Wade's rules were extended from boranes to Zintl structures.

Let us examine some transformations of the *closo*-boranes, $B_nH_n^{2-}$

$$B_nH_n^{2-} \rightarrow (:B)_n^{-n-2} + nH^+$$
$$(:B)_n^{-n-2} \xrightarrow{2p} (:Bi)_n^{n-2}$$

1. In the first step, the terminal B—H bonds are cleaved heterolytically to give *exo*-polyhedral lone pairs on the borane.
2. The second step occurs by inserting two protons into each nucleus thereby reducing the charge and converting to Triel (Tr, group 13) B to a pnictogen (Pn, group 15) cluster, namely that of Bi. The important thing is that during this imagined transformation the intracage bonding does not change.
3. Therefore, the rules developed to rationalize borane clusters (Wade's rules) can be applied to main group clusters, such as those found in the Zintl ions (see following section).

3.1 Zintl anions

Zintl anions are named after the German chemist Eduard Zintl who pioneered their study in the 1920s and 1930s.

1. They are formed when very electropositive elements (group 1 or 2) react with moderately electronegative elements (P, As, Sb, Se), metalloid (Ge, Ga), or metallic elements (Tl, Sn, Pb). In these compounds, there is essentially a complete transfer of electrons from the electropositive element to the main group elements, forming negatively charged polyhedra, chains, or sheets.

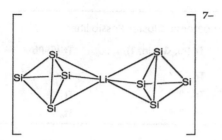

■ **FIGURE 9.9** Zintl structure of the anion within K_7LiSi_8.

2. The cluster K_7LiSi_8 is an example of a Zintl anion, with the structure being shown in Fig. 9.9. This can be viewed as two $Si_4{}^{4-}$ units bridged by a Li^+. Each $Si_4{}^{4-}$ has three electrons for cage bonding (each five-electron Si^- would have an *exo*-polyhedral lone pair and three cage electrons) giving a total of 12 electrons, or $2n + 2$ pairs for a *nido*-structure. In this regard, the tetrahedral structure can be considered as being derived by the removal of one vertex from a trigonal bipyramidal *closo*-cage.

3. Fig. 9.10 shows some group 14 Zintl structures in which $[Sn_5]^{2-}$ (or $[Pb_5]^{2-}$) would be a 22-electron system with 5 *exo*-polyhedral electron pairs and 12 $(n + 1)$ pairs for cage bonding to give a *closo*-structure. Note that $[Ge_9]^{2-}$ has 10 electron pairs $(n + 1)$ for a *closo*-structure, whereas $[Ge_9]^{3-}$ and $[Sn_9]^{3-}$ do not. Evidently, the extra electron does not destabilize the *closo*-structure sufficiently to produce a *nido*-cage. However, two extra electrons, as in $[Ge_9]^{4-}$, lead to the expected *nido*-structure.

4. Table 9.4 gives the possibilities for the Triel (Tr, group 13), tetrel (Tt, group 14), and pnictogen (Pn, group 15) families.

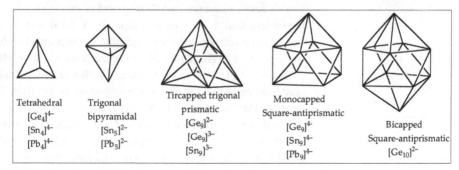

| Tetrahedral $[Ge_4]^{4-}$ $[Sn_4]^{4-}$ $[Pb_4]^{4-}$ | Trigonal bipyramidal $[Sn_5]^{2-}$ $[Pb_5]^{2-}$ | Tricapped trigonal prismatic $[Ge_9]^{2-}$ $[Ge_9]^{3-}$ $[Sn_9]^{3-}$ | Monocapped Square-antiprismatic $[Ge_9]^{4-}$ $[Sn_9]^{4-}$ $[Pb_9]^{4-}$ | Bicapped Square-antiprismatic $[Ge_{10}]^{2-}$ |

■ **FIGURE 9.10** Zintl structures of group 14 elements.

Table 9.4 Homoatomic Cluster Possibilities

	Tr (Al, Ga, In, Tl)	Tt (Si-Pb)	Pn (P-Bi)
closo	Tr_n^{-n-2}	Tt_n^{2-}	Pn_n^{n-2}
nido	Tr_n^{-n-4}	Tt_n^{4-}	Pn_n^{n-4}
arachno	Tr_n^{-n-6}	Tt_n^{6-}	Pn_n^{n-6}

Thus, Se_4^{2+} and Sb_4^{2-} are 22-electron systems having 14 cage electrons (7 or $n + 3$ electron pairs) for an *arachno*-structure, which is consistent with their square planar geometry.

3.2 Other main group cages

1. The rules can be extended to rationalize other structures. White phosphorus, P_4, is a 20-valence electron system with 12 electrons for cage bonding, or $n + 2$ electron pairs (6 pairs $= 4 + 2$) for a *nido*-cage. This is consistent with a tetrahedral structure that can be derived by the removal of one vertex from a *closo*-trigonal bipyramidal cage. Tetrahedrane, C_4H_4, can be similarly viewed.
2. In the same way, the electron count of benzene (C_6H_6) (nine pairs for cage bonding or $n + 3$ pairs) could be viewed as an *arachno*-cage derived by removing opposing vertices of a hexagonal bipyramidal cage.
3. These compounds are electron precise and are more conveniently described using standard $2c$-$2e$ electron pair bonds.

4. EXTENSION OF WADE'S RULES BEYOND BORON CLUSTERS

1. Electron counting rules—polyhedral electron count (pec). The basis for Wade's rules constitutes an MO treatment of the cages.
2. If you examine a *closo*-cage composed of fused deltahedra with n vertices, you will see that each cage atom has four atomic orbitals (s, p_x, p_y, p_z) that give a total of $4n$ orbitals from which $4n$ MOs can be constructed. Calculations show that $2n - 1$ MOs are unavailable for cage bonding as they are either antibonding or are not properly directed. That leaves $2n + 1$ MOs for cluster bonding. Of these, n are involved in *exo*-polyhedral bonding (that is, directed out of the cage) leaving $n + 1$ MOs for cage bonding, which could accommodate $n + 1$ electron pairs ($2n + 2$ electrons). The total number of valence electrons (pec) would be $4n + 2$ for a stable *closo*-cage.

Table 9.5 Summary of Closed-Shell Requirements for the Electrons of Main Group and Transition Metal *n*-Vertex Fused Deltahedra Cages

	Main Group MO's				Transition Metal MO's			
Structure	Total	Unavailable	Available	pec[a]	Total	Unavailable	Available	pec[a]
closo	$4n$	$2n-1$	$2n+1$	$4n+2$	$9n$	$2n-1$	$7n+1$	$14n+2$
nido	$4n$	$2n-2$	$2n+2$	$4n+4$	$9n$	$2n-2$	$7n+2$	$14n+4$
arachno	$4n$	$2n-3$	$2n+3$	$4n+6$	$9n$	$2n-3$	$7n+3$	$14n+6$

[a]Polyhedral electron count.

3. In the same way, for a *nido*-cage of *n* vertices, the $4n$ MOs would be partitioned between $2n-2$ unavailable MOs and $2n+2$ cluster MOs of which n are *exo*-polyhedral and $n+2$ are involved in cage bonding. Therefore, a stable *nido*-cluster should be one with $4n+4$ valence electrons (pec $= 4n+4$). Table 9.5 summarizes this argument and Fig. 9.11 shows some examples.

4. For the main group elements, Table 9.5 is just another way of presenting the information in Table 9.1 (Wade's rules) in that the total number of cluster electrons (pec) is given in Table 9.5, whereas Table 9.1 concentrates on the number of electron pairs involved in cage bonding.

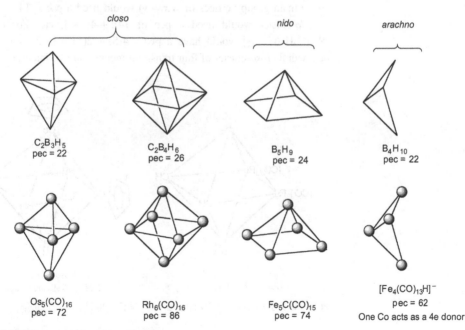

	closo	*nido*	*arachno*

$C_2B_3H_5$
pec $= 22$

$C_2B_4H_6$
pec $= 26$

B_5H_9
pec $= 24$

B_4H_{10}
pec $= 22$

$Os_5(CO)_{16}$
pec $= 72$

$Rh_6(CO)_{16}$
pec $= 86$

$Fe_5C(CO)_{15}$
pec $= 74$

$[Fe_4(CO)_{13}H]^-$
pec $= 62$
One Co acts as a 4e donor

■ **FIGURE 9.11** Main group and transition metal clusters.

5. The data in Table 9.5 show that in the MO analysis of the structures of some organometallic carbonyl clusters, the number of unavailable MOs is the same as that found for main group clusters. This allows the extension of the pec's to organometallic clusters. Fig. 9.11 shows some examples.

6. Note that in obtaining the pec the relationship is pec = No. of valence electrons on the metal + Number contributed by the ligands. The last is obtained assuming standard ligand electron contributions. That is, CO, CN^-, NO^+, X^-, etc., are two-electron donors; $[R_5C_5]^-$, C_6H_6 are six-electron donors, C_5H_5 is a five-electron donor, etc.

7. Two transition metal compounds as examples:
   $Os_5(CO)_{16}$ pec = $5(8) + 16(2) = 72 = 14(5) + 2$ or $14n + 2$ for a *closo*-cluster.
   $[Ru_6(CO)_{18}]^{2-}$ pec = $6(8) + 18(2) + 2 = 86 = 14(6) + 2$ or $14n + 2$ for a *closo*-cluster.

4.1 **Mixed main group/transition metal clusters**

For all transition metal complexes or all main group complexes, the pec's for the different types of clusters are given in Table 9.5. The counting for a complex in which both the main group and transition metal groups occupy the vertices can be obtained by combining the two counting rules.

A *closo*-compound with n vertices composed of x transition metal vertices and y main group vertices ($n = x + y$) would need a pec = $14x + 4y + 2$, a *nido*-cluster would need a pec of $14x + 4y + 4$, etc. For example, $[Ru_4C_2H_2(CO)_{12}]$ would have a pec = $4(8) + 2(5) + 12(2) = 66$. For an octahedral *closo*-cluster of four transition metals and two main group atoms

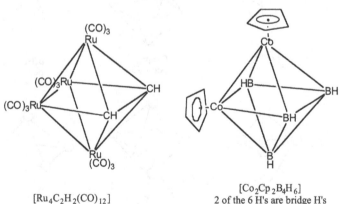

$[Ru_4C_2H_2(CO)_{12}]$

$[Co_2Cp_2B_4H_6]$
2 of the 6 H's are bridge H's

■ **FIGURE 9.12** Experimentally determined structures of $[Ru_4C_2H_2(CO)_{12}]$ and $Co_2Cp_2B_4H_6$.

Table 9.6 Number of Contributing Electrons for Cluster Bonding

Cluster Fragment	Group 6 Cr, Mo, W	Group 7 Mn, Tc, Re	Group 8 Fe, Ru, Os	Group 9 Co, Rh, Ir
$M(CO)_2$	−2	−1	0	1
$M(CO)_3$	0	1	2	3
$M(CO)_4$	2	3	4	5
$M(Cp)$	−1	0	1	2
$M(C_6H_6)$	0	1	2	3

one would need a pec $= 14(4) + 2(4) + 2 = 66$. The structure of this complex, shown in Fig. 9.12, was determined by X-ray crystallography. Similarly, $Co_2Cp_2B_4H_6$ has a pec $= 2(9) + 2(5) + 4(3) + 6(1) = 6$. For a *closo*-six vertex cluster of two transition metals and four main group atoms the pec should be $= 14(2) + 4(4) + 2 = 46$. The structure of this cluster is shown in Fig. 9.12.

For the mixed clusters, it is sometimes simpler to look at electron pairs used in cluster, as shown in Table 9.1. In doing so, the electrons used in cluster bonding (E) for some common organometallic groups are given in Table 9.6 (E = pec − 12).

In $(CoCp)_2B_4H_6$ cluster, the electron counting will be E $= 2(2) + 4(2) + 2 = 14 = 7$ pairs $= n + 1$ counting for a *closo*-cluster.

4.2 Capping groups

Some compounds will have an electron count lower than that for a *closo*-structure, but such complexes will have a *closo*-cage with capping groups on one or more of the triangular faces. The capping groups do not require any additional cluster electrons, since they are technically "outside" of the cluster. Therefore, a cluster with n vertices of which one is capping will have n electron pairs involved in cluster bonding to have a pec of $4n$ for main group atom clusters or $14n$ for transition metal clusters. If two are capping, then the pec requirement goes down by another two as in $Os_6(CO)_{18}$ with pec $= 6(8) + 18(2) = 84 = 14(6) = 14n$. The structure is that of a five-vertex *closo*-cage with an $Os(CO)_3$ capping one of the faces. On the other hand, $[Os_8(CO)_{22}]^{2-}$ will have a pec $= 8(8) + 22(2) + 2 = 110 = 14(8)−2$. Therefore, the structure would be that of a six-vertex octahedron with two capping groups. These two structures are shown in Fig. 9.13.

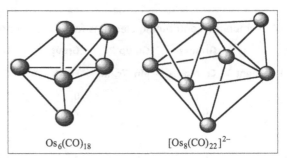

$Os_6(CO)_{18}$ $[Os_8(CO)_{22}]^{2-}$

■ **FIGURE 9.13** Metal atom positions for two capped osmium carbonyl clusters.

It should be noted that the $Os_6(CO)_{18}$ structure could be described as a bicapped tetrahedron. The description as a monocapped trigonal bipyramid is more in line with the pec of 84.

4.3 Condensed clusters

One can view a monocapped cluster as resulting from two clusters condensing to share a common face. For example, the $Os_6(CO)_{18}$ could be viewed as arising from a trigonal bipyramidal and a tetrahedral cluster sharing a common triangular face, as shown in Fig. 9.14.

The reason this view is useful is that there are other examples of condensed clusters in which atoms and edges are shared. The pec for a condensed cluster is equal to the total number required for the separate clusters minus the number associated with the shared unit.

The numbers of electrons to be subtracted are: 18 for a shared atom, 34 for a shared edge, and 48 for a shared face. There is a measure of ambiguity in how to describe some of the basic cage shapes. For example, the tetrahedron can be viewed as being either a four-vertex *closo*-cluster (pec = 58) or a *nido*-cluster (pec = 60), derived by removing a vertex from a trigonal bipyr-amidal cage. The X-ray structure of $Rh_4(CO)_{12}$ (pec = 60) is that of a tetra-hedron, which is consistent with a *nido*-description. Fig. 9.15 shows some clusters with the expected pec for each.

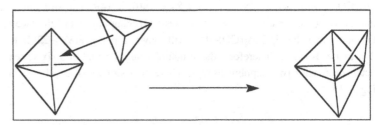

■ **FIGURE 9.14** Fusion of a trigonal bipyramid and tetrahedral cluster.

Cluster framework	Cage	pec
Triangular	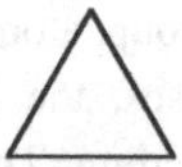	48
Tetrahedron	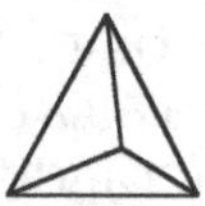	60
Buterfly or planar		62
square		64
Trigonal prism		90

■ **FIGURE 9.15** Structures and corresponding pec's.

Using these counts, the formulas for the fused cages shown in Fig. 9.16 can be rationalized.

The pec for $Os_5(CO)_{19} = 78$; for two triangular clusters sharing a vertex, the pec should be $= 2(48) - 18 = 78$. The pec for $H_2Os_5(CO)_{16} = 60 + 48 - 34 = 74$. Similarly, the pec for $H_2Os_6(CO)_{18} = 74 + 60 - 48 = 86$.

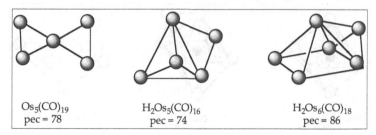

$Os_5(CO)_{19}$	$H_2Os_5(CO)_{16}$	$H_2Os_6(CO)_{18}$
pec = 78	pec = 74	pec = 86

■ **FIGURE 9.16** Structures of some fused cage clusters.

4.4 **Clusters with interstitial atoms**

There are a number of examples of transition metal clusters that contain a main group atom in semi- or fully interstitial positions in the cage. In such cases, the interstitial atom contributes all its valence electrons to cage bonding. Three examples are shown in Fig. 9.17.

1. $Ru_6(CO)_{17}C$. The $pec = 6(8) + 17(2) + 4 = 86 = 14(6) + 2$ or $14n + 2$ for a *closo*-cage.
2. $[Re_7(CO)_{22}C]^-$. The $pec = 7(7) + 22(2) + 4 + 1 = 98 = 14(7) = 14n$ for a monocapped octahedral cluster.
3. $[Fe_4(CO)_{12}C]^{2-}$. The $pec = 4(8) + 12(2) + 4 + 2 = 62 = pec$ for an *arachno*-butterfly cluster.

4.5 **Isolobal relationships**

The extension of Wade's rules to main group and transition metal clusters is based on the fact that all use orbitals with similar lobal characteristics in forming the particular cages and are isolobal with boron. The use of isolobal relationships of molecular fragments can be useful in predicting the viability and structures of a number of molecules other than clusters.

In constructing metal-ligand bonds, a main group element can use its ns and np orbitals or combinations of such orbitals in hybrids, as shown in Fig. 9.18.

For transition metal complexes in which the metal conforms to the noble gas rule or effective atomic number (EAN) rule, some of the metal's electrons will be in nonbonding orbitals (either atomic orbitals or hybrid orbitals pointing away from the ligands). Table 9.7 presents the predicted most common geometries for coordination numbers of 4–7, along with the number of nonbonding orbitals. In general, the bonding orbitals will hybridize to give the strongest possible bonds. One can derive molecular fragments by

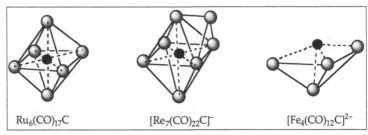

| $Ru_6(CO)_{17}C$ | $[Re_7(CO)_{22}C]^-$ | $[Fe_4(CO)_{12}C]^{2-}$ |

■ **FIGURE 9.17** Three interstitial clusters.

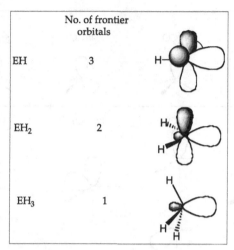

	No. of frontier orbitals	
EH	3	H—
EH$_2$	2	H
EH$_3$	1	H

■ **FIGURE 9.18** Bonding possibilities in main group fragments.

Table 9.7 Valence Orbitals for Some Common Geometries

Coordination Number	Geometry	Hybrid Schemes	Nonbonding Orbitals
4	Tetrahedral	sp^3 or sd^3	5
5	Trigonal bipyramid	sp^3d or spd^3	4
6	Octahedral	d^2sp^3	3
6	Trigonal prism	d^2sp^3 or spd^4	3
7	Pentagonal bipyramid	sp^3d^3	2

removing one or more adjacent ligands, leaving behind the out-pointing hybrid orbitals. This is illustrated in Fig. 9.19.

To the first approximation, the orbital characteristics of each fragment are similar. For example, $Ni(CO)_3$, $Fe(CO)_4$, $Cr(CO)_5$, and $M(CO)_7$ are similar to CH_3^+ in that they all have one outwardly pointing orbital, and are, hence, isolobal. The symbol $\longleftrightarrow$ is used to indicate isolobal fragments.

The following are some isolobal relationships:

$$Ni(CO)_3 \longleftrightarrow Fe(CO)_4 \longleftrightarrow Cr(CO)_5 \longleftrightarrow Ti(CO)_6 \longleftrightarrow CH_3^+$$

The presence of an additional electron in the outwardly pointing orbital would give rise to $Cu(CO)_3 \longleftrightarrow Co(CO)_4 \longleftrightarrow Mn(CO)_5 \longleftrightarrow V(CO)_6 \longleftrightarrow CH_3$. Other isolobal fragments are given and some resulting structures are shown in Fig. 9.20.

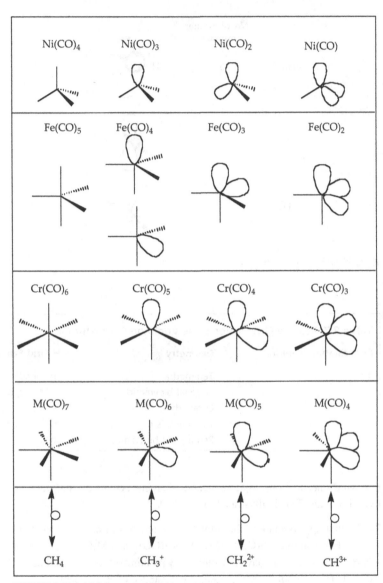

■ **FIGURE 9.19** Isolobal fragments of transition metal carbonyls.

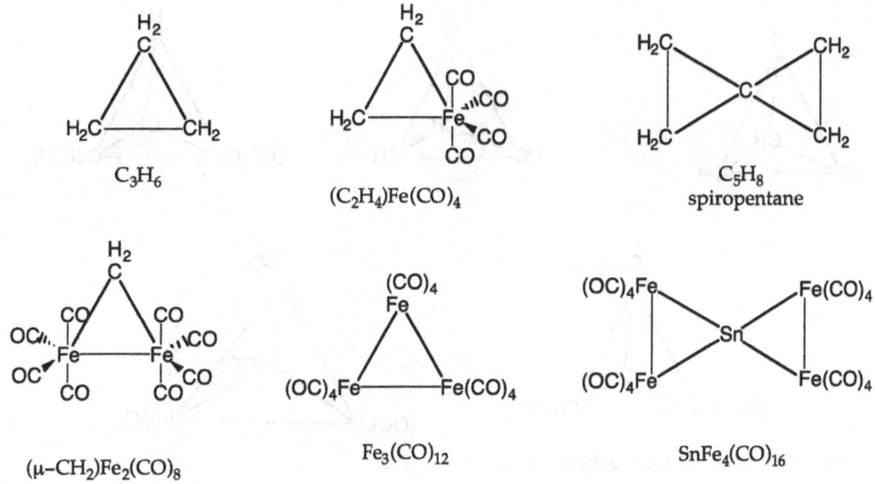

C3H6

(C2H4)Fe(CO)4

C5H8
spiropentane

(μ–CH2)Fe2(CO)8

Fe3(CO)12

SnFe4(CO)16

$Fe(CO)_4 \longleftrightarrow CH_2$

■ **FIGURE 9.20** Structures showing isolobal relationship between Fe(CO)$_4$ and CH$_2$ units.

$Ni(CO)_2 \longleftrightarrow Fe(CO)_3 \longleftrightarrow Cr(CO)_4 \longleftrightarrow Ti(CO)_5 \longleftrightarrow CH_2^{2+}$

$Cu(CO)_2 \longleftrightarrow Co(CO)_3 \longleftrightarrow Mn(CO)_4 \longleftrightarrow V(CO)_5 \longleftrightarrow CH_2^{+}$

$Ni(CO)_3 \longleftrightarrow Fe(CO)_4 \longleftrightarrow Cr(CO)_5 \longleftrightarrow CH_2$

$Ni(CO) \longleftrightarrow Fe(CO)_2 \longleftrightarrow Cr(CO)_3 \longleftrightarrow Ti(CO)_4 \longleftrightarrow CH^{3+}$

$Cu(CO) \longleftrightarrow Co(CO)_2 \longleftrightarrow Mn(CO)_3 \longleftrightarrow V(CO)_4 \longleftrightarrow CH^{2+}$

$Ni(CO)_2 \longleftrightarrow Fe(CO)_3 \longleftrightarrow Cr(CO)_4 \longleftrightarrow CH^{+}$

$Co(CO)_3 \longleftrightarrow Mn(CO)_4 \longleftrightarrow CH$

These isolobal relationships can be used to rationalize the following progressions. In Fig. 9.20, the Fe(CO)$_4$ fragment, which is isolobal with CH$_2$, is the one with two out-pointing lobes derived from an octahedral and not the one derived from the trigonal bipyramid.

In the same way, the three out-pointing Co(CO)$_3$ fragments, which are isolobal to CH, can be used for inferring organometallic molecules derived from tetrahedrane C$_4$H$_4$ (Fig. 9.21).

Isolobal relationships can be worked out for other noble gas structures. For example, $(\eta^5\text{-C}_5\text{H}_5)\text{Co(CO)}_2$ is an 18-electron compound (Co = 9,

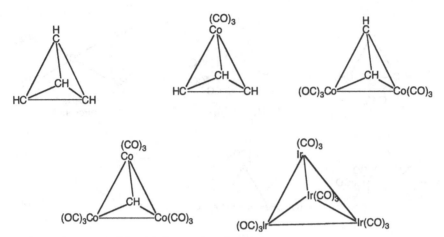

■ **FIGURE 9.21** Isolobal relationship between $M(CO)_3$ and CH fragments (M = Co or Ir).

$C_5H_5 = 5$, $2CO = 4$, total $= 18$). Therefore the following isolobal relationships, shown in Fig. 9.22, can be envisioned.

Using the same progression as used for the metal carbonyl fragments, the following isolobal relationship can be written:

$$CH^{2+} \longleftrightarrow (C_5H_5)Fe; CH^{+} \longleftrightarrow (C_5H_5)Co; CH \longleftrightarrow (C_5H_5)Ni.$$

Also note that CH^{+} is isoelectronic and isolobal with a BH fragment. Therefore the compound, shown in Fig. 9.23, will have a cage bonding electron

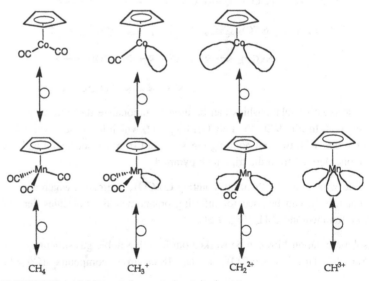

■ **FIGURE 9.22** Isolobal fragments for some cyclopentadienyl moieties.

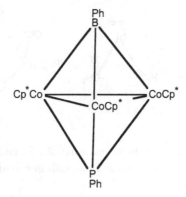

■ **FIGURE 9.23** Five-vertex *closo*-cage structure of [((CoCp*)₃B(C₆H₅)P(C₆H₅)].

count of 12 (two each from the BPh and the three CoCp* and four from the PPh) or six electron pairs $(n + 1)$ for a five-vertex *closo*-cage structure.

4.6 **Isolobal relationships with fragments not derived from noble gas structures**

1. The aforementioned relationships were derived from fragments of complexes that originally had noble gas electron counts (18 for transition metals and 8 for main group elements). The square planar d^8 metals (Pt(II), Pd(II), Au(III), Ir(I), Rh(I)) form square planar complexes with 16 valence electrons, rather than 18-electron five coordinate complexes.
2. The following isolobal relationships, shown in Fig. 9.24, can be written for fragments derived from square planar d^8 complexes.

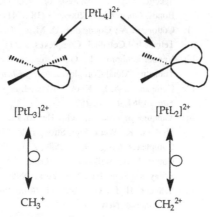

■ **FIGURE 9.24** Isolobal relationship for angular and T-shaped [PtLₙ]²⁺ fragments.

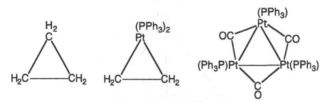

■ **FIGURE 9.25** Structures showing the isolobal relationship between PtL_2 and CH_2 fragments.

3. From these, the isolobal relationship $PtL_2 \longleftrightarrow CH_2$ can be derived. This leads to a rationalization of the progression shown in Fig. 9.25.

BIBLIOGRAPHY

1. Morrill, T. C.; Silverstein, R. M.; Bassler, G. C. Spectrometric Identification of Organic Compounds. *J. Chem. Educ.* **1963,** *39* (11), 546.
2. Kagan, M. R.; Guilbault, G. G. Ion-mobility Measurements of Inorganic and Organic Phosphorus Compounds. *J. Phys. Chem.* **1968,** *72* (8).
3. Crouch, S.; Skoog, D. A. *Principles of Instrumental Analysis;* Thomson Brooks/Cole: Australia, 2007; pp 335–398.
4. Herzberg, G. Molecular Spectra and Molecular Structure. *J. Chem. Educ.* **1950**.
5. DeVries, J. Valence and Molecular Structure. *J. Chem. Educ.* **1936,** *13* (7), 320.
6. Bowman, D.; Jakubikova, E. Low-Spin versus High-spin in Pseudo-octahedral Iron Complexes. *Inorg. Chem.* **2012,** *51* (11), 6011–6019.
7. Jones, G. R. H. Substitution Reactions in Octahedral Complexes. *J. Chem. Educ.* **1966,** *43* (12), 657.
8. Kettle, S. F. A. Ligand Group Orbitals of Octahedral Complexes. *J. Chem. Educ.* **1966,** *43* (1), 21.
9. Cotton, F. A.; Goodgame, D. M. L. New Tetrahedral Complexes of Nickel(II). *J. Am. Chem. Soc.* **1960,** *82* (22), 5771–5774.
10. Mehn, M. P.; Brown, S. D.; Jenkins, D. M.; Peters, J. C.; Que, L. Vibrational Spectroscopy and Analysis of Pseudo-tetrahedral Complexes with Metal Imido Bonds. *Inorg. Chem.* **2006,** *45* (18), 7417–7427.
11. Cotton, F. A.; Goodgame, D. M. L.; Goodgame, M. The Electronic Structures of Tetrahedral Cobalt(II) Complexes. *J. Am. Chem. Soc.* **1961,** *82* (23), 4690–4699.
12. Vanquickenborne, L. G.; Verdonck, E. Charge-transfer Spectra of Tetrahedral Transition Metal Complexes. *Inorg. Chem.* **1976,** *15* (2), 454–461.
13. Companion, A. L.; Komarynsky, M. A. Crystal Field Splitting Diagrams. *J. Chem. Educ.* **1964,** *41* (5), 257.
14. Vanquickenborne, L. G.; Hendrickx, M.; Postelmans, D.; Hyla-Kryspin, I.; Pierloot, K. Weak-field Strong-field Correlation Diagrams in Transition Metal Complexes. *Inorg. Chem.* **1988,** *27* (5), 900–907.
15. Cotton, F. A.; Wilkinson, G.; Gaus, P. L. *Basic Inorganic Chemistry,* 3rd ed.; John Wiley and Sons, Inc.: New York, 1995.
16. Schlafer, H. L.; Gliemann, G. *Basic Principles of Ligand Field Theory;* Wiley-Interscience: New York, 1969.

17. Lever, A. P. B. *Inorganic Electronic Spectroscopy,* 2nd ed.; Elsevier Publishing Co.: Amsterdam, 1984.
18. Seltzer, M. D. Interpretation of the Emission Spectra of Trivalent Chromium-doped Garnet Crystals Using Tanabe-sugano Diagrams. *J. Chem. Educ.* **1995,** *72* (10), 886.
19. Lamonova, K. V.; Zhitlukhina, E. S.; Babkin, R. Y.; Orel, S. M.; Ovchinnikov, S. G.; Pashkevich, Y. G. Intermediate-Spin State of a 3d Ion in the Octahedral Environment and Generalization of the Tanabe-sugano Diagrams. *J. Phys. Chem.* **2011,** *115* (46), 13596−13604.
20. Lever, A. P. B. Charge Transfer Spectra of Transition Metal Complexes. *J. Chem. Educ.* **1974,** *51* (9), 612.

Part 4

Advanced Topics-3: Organometallic Chemistry and Catalysis

Chapter

10

Organometallic Chemistry

1. INTRODUCTION: WHAT IS IN ORGANOMETALLIC CHEMISTRY?

Coordination chemistry is the branch of science that deals with the interactions of organic and inorganic ligands involving metal centers, whereas organometallic chemistry involves chemical complexes in which at least one bond is present between a metal and a carbon atom of an organic species. In other words, this area of chemistry lies at the interface between organic chemistry and the inorganic chemistry of metals. The development of organometallic chemistry was inevitable because many organic reactions are unable to proceed under normal conditions without the presence of a metal or a metal catalyst. For example, our body chemistry cannot function without essential enzymes that involve metal-bound proteins. Another example would be the production of organic polymers under suitable catalytic conditions. This involves organometallic compounds that utilize a vacant site on their metal center to undergo oxidative addition and reductive elimination cycles. Other than the transition metals, lithium and magnesium play an important role within ionic functional moieties, such as Grignard reagents (RMgX), butyllithium (BuLi), and the Reformatsky reagent (RZnX), with enhancement of their nucleophilic character. However, most organometallic reagents are sensitive to air and moisture and, therefore, they must be handled in an inert atmosphere. Many pharmaceutical industries have adopted catalytic processes involving organometallic complexes in the syntheses of new effective drugs. Several Nobel Prize winners, working with catalytic reactions, invented Ziegler–Natta polymerization, Suzuki–Sonogashira couplings, Click Chemistry, Schrock/Grubbs/Chauvin olefin metathesis, etc. After learning about the behavior of organometallic compounds and their involvement in reactions that affect almost every aspect of our life, you will have the satisfaction and fulfillment of learning not only inorganic and organic chemistry but also the combined discipline called organometallic chemistry, which has emerged in the last few decades.

Advanced Inorganic Chemistry. http://dx.doi.org/10.1016/B978-0-12-801982-5.00010-2

2. DEFINITIONS AND NOMENCLATURE OF ORGANOMETALLIC COMPOUNDS

2.1 Types of organometallic compounds

Transition metal complexes where the ligands are CO, alkenes, alkynes, aromatics, or allyl radicals are called organometallic compounds.

For the π–type ligands, the primary metal–ligand bond is formed by the overlap of a vacant metal orbital with a filled ligand π orbital.

The transition metal is in its low oxidation state.

There is extensive back π-bonding between the metal and the ligand.

The complexes are low-spin complexes in which maximum spin pairing occurs. Allyl radicals are odd electron π systems and act as odd electron donors.

The two most common donors are:

1. Allyl radical $CH_2\!\!=\!\!CH\!-\!CH_2{}^\bullet$ (a three-electron donor) and
2. Cyclopentadienyl $C_5H_5{}^\bullet$ (a five-electron donor)

These donor radical ligands, in which one electron is in each p_π orbital, are shown in Fig. 10.1.

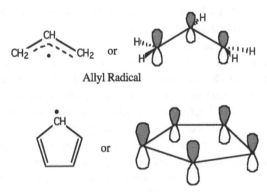

Allyl Radical

Cyclopentadienyl Radical

■ **FIGURE 10.1** Three- and five-electron donor radical ligands.

2.2 IUPAC nomenclature for organometallic compounds

1. Normal IUPAC (International Union of Pure and Applied Chemistry) rules of nomenclature hold. The only thing new is how to name the π ligands.
2. The name of the ligand is the name of the organic molecule (or radical) with the prefix η (η is read as eta or hapto, from the Greek meaning to fasten). The η is raised to a power that gives the number of carbon atoms to which the metal is bonded.
3. Examples are: $[PtCl_2(NH_3)(C_2H_4)]$ is named as amminedichloro (η^2-ethene)-platinum(II) (Zeise's salt) and $[ReH(C_5H_5)_2]$ is named as bis(η^5-cyclopentedienyl)-hydridorhenium(I).
4. If the ligand does not use all the π-bonding, carbon atoms of an unsaturated chain in attaching to the metal, location numbers are inserted preceding the η. An example is shown in Fig. 10.2.

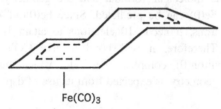

Fe(CO)₃

■ **FIGURE 10.2** Tricarbonyl(1-4-η^4-cyclooctatetraene)iron(0).

3. MOLECULAR FORMULAS AND STRUCTURES OF ORGANOMETALLIC COMPOUNDS

3.1 Effective atomic number rule

1. The most stable organometallics are those in which the transition metal utilizes all of its valence d, s, and p orbitals in bonding. In these complexes the transition metal will have 18 valence electrons or have an effective atomic number (EAN) = 18.
2. The transition metal will be in its lowest spin state and most complexes are diamagnetic.

3.2 Metal carbonyls

1. In most metal carbonyls, the metal is in a zero oxidation state and the EAN rule is obeyed and bonding by the CO is through the lone pair of electrons on the carbon atom ($\leftarrow |C{\equiv}O|$).

2. If the metal has an even number of electrons, a simple carbonyl is formed. For example, carbonyl complexes of Ni(0) can be predicted by knowing that Ni($3d^8 4s^2$) has 10 valence electrons. Therefore, it needs eight electrons to attain 18 to obey the EAN Rule. Thus, it will bond to four CO's to form a tetracarbonylnickel(0) complex, $Ni(CO)_4$, which is a tetrahedral molecule as expected from its sp^3 hybridization indicated in Fig. 10.3.

```
                    3d              4s        4p
Ni(0)  ↑↓  ↑↓  ↑↓  ↑↓  ↑↓          xx     xx  xx  xx
                                   CO     CO  CO  CO
                                   |---sp³ hybridized---|
```

■ **FIGURE 10.3** Electronic arrangement of $Ni(CO)_4$.

Similarly, the molecular formula and the geometry for carbonyl complex of Fe(0) can be predicted. Since Fe(0)($3d^6 4s^2$) has eight valence electrons, it needs 10 electrons to attain 18 to obey the EAN rule. Therefore, it will bond to five CO's to form the pentacarbonyliron(0) complex, $Fe(CO)_5$, which has a trigonal bipyramidal geometry as expected from its use of dsp^3 hybridization (Fig. 10.4).

```
                  3d                 4s        4p
Fe(0)  ↑↓  ↑↓  ↑↓  ↑↓  xx           xx     xx  xx  xx
                       CO           CO     CO  CO  CO
                       |--------dsp³ hybridized-----|
```

■ **FIGURE 10.4** Electronic arrangement of $Fe(CO)_5$.

3. If the metal has an odd number of electrons, dimeric carbonyls are formed. An example can be found for the carbonyl complex of cobalt in its zero oxidation state. Since Co(0) ($3d^7 4s^2$) has nine valence electrons, it needs nine electrons to attain 18 to obey the EAN rule. Therefore, each of the two Co's will bond to 4CO's and then dimerize to form octacarbonyldicobalt(0), $Co_2(CO)_8$, which is a diamagnetic complex due to the formation of Co—Co metal bond with dsp^3 hybridization for each cobalt metal (Fig. 10.5).

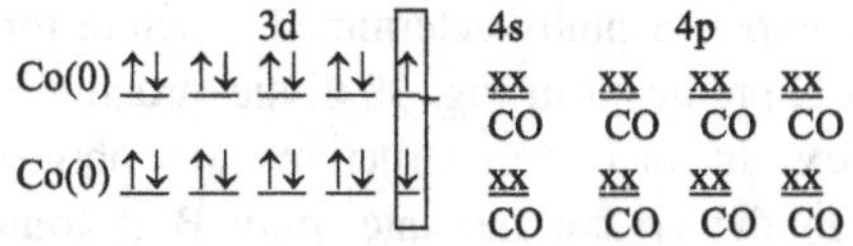

3d 4s 4p

Co(0) ↑↓ ↑↓ ↑↓ ↑↓ ↑ | XX XX XX XX
CO CO CO CO

Co(0) ↑↓ ↑↓ ↑↓ ↑↓ ↓ | XX XX XX XX
CO CO CO CO

■ **FIGURE 10.5** Electronic arrangement of $Co_2(CO)_8$.

Similarly, the molecular formula and the corresponding geometry for carbonyl complex of Mn(0) can be predicted. Since Mn(0) $(3d^5 4s^2)$ has seven valence electrons, it needs 11 more to attain the 18-electron system to obey the EAN rule. Therefore, two Mn's are required where each will bond to five CO's and then dimerize to form diamagnetic decacarbonyldimanganese(0), $Mn_2(CO)_{10}$, with formation of a Mn–Mn bond.

4. The above rules hold for the simplest metal carbonyls except for the very early transition metals. For example, vanadium in its zero oxidation state $(3d^3 4s^2)$ has five valence electrons and it needs 13 to obey the EAN rule. Consequently, dimerization of two $V(CO)_6$ can be expected. However, due to the large number of CO ligands it is not sterically possible. Therefore the simplest vanadium carbonyl is $V(CO)_6$, which does not obey the EAN rule. Compared with the other metal carbonyls, $V(CO)_6$ is quite unstable and it is easily reduced. The resulting $[V(CO)_6]^-$ anion obeys the EAN rule and, consequently, it is very stable.

$$V(CO)_6 + Na \rightarrow Na^+ \left[V(CO)_6\right]^-$$

Although the EAN rule can help in predicting the formulas of the simplest metal carbonyl complexes, it is of limited use in predicting the detailed structures. For example, the electronic arrangement in $Mn_2(CO)_{10}$ is such that each Mn is bonded to five CO's with one Mn–Mn bond, leading to d^2sp^3 hybridization, and the structure will be with two octahedra sharing a vertex as shown in Fig. 10.6.

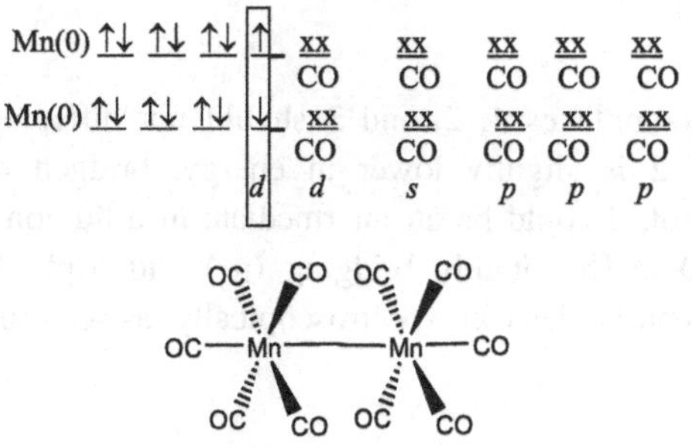

Mn(0) ↑↓ ↑↓ ↑↓ ↑ | XX XX XX XX XX
CO CO CO CO CO

Mn(0) ↑↓ ↑↓ ↑↓ ↓ | XX XX XX XX XX
CO CO CO CO CO
d *d* *s* *p* *p* *p*

■ **FIGURE 10.6** Electronic arrangement and structure of $Mn_2(CO)_{10}$.

Although the experimentally determined structure for $Mn_2(CO)_{10}$ is exactly what is predicted in Fig. 10.6, the structure of $Co_2(CO)_8$ is more complex. In fact, two structures are observed, A and B (Fig. 10.7). In the crystalline state, only B is found, whereas in solution, both A and B exist. The structure of $Co_2(CO)_8$ is an example where a bridging carbonyl exists. This is an important aspect of metal carbonyl chemistry, from an electron counting perspective where both A and B are acceptable (a terminal CO is formally a two-electron donor, whereas each bridging carbonyl can be thought of as contributing one electron to each metal).

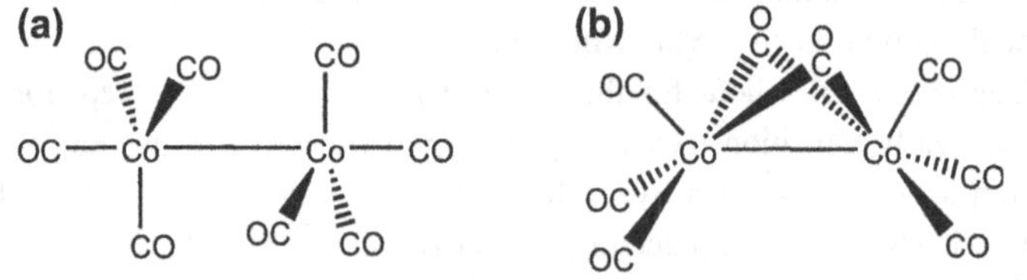

■ **FIGURE 10.7** Possible two structures of $Co_2(CO)_8$.

These results suggested that the process shown in Fig. 10.8 is possible, and frequently observed, in multimetal carbonyl complex.

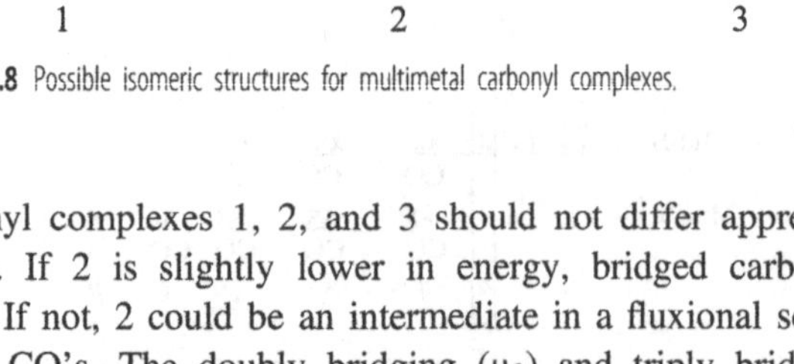

■ **FIGURE 10.8** Possible isomeric structures for multimetal carbonyl complexes.

Carbonyl complexes 1, 2, and 3 should not differ appreciably in energy. If 2 is slightly lower in energy, bridged carbonyls are found. If not, 2 could be an intermediate in a fluxional scrambling of the CO's. The doubly bridging (μ_2) and triply bridging (μ_3) carbonyls can be detected spectroscopically, as seen in Table 10.1.

Table 10.1 Infrared Stretching Energies for Neutral Carbonyls

Type of CO	Energy (cm^{-1})
Free CO	2143
Terminal	1850–2125
μ_2-bridging	1700–1860
μ_3-bridging	1600–1700

The change in stretching energy can be understood if we recall that the CO is a good back π-bonding ligand and both ligand-to-metal and metal-to-ligand electron transfers are important. The former process would tend to increase the vibrational energy, whereas the latter would weaken the C—O bond by populating the ligand's π^* antibonding molecular orbitals. The more the number of metals involved in this back π-bonding, the lower will be the C—O stretching energy. As one would expect, anionic carbonyls should exhibit decreased CO stretching energies and cationic ones, increased energies. For example, the CO stretching energy in $[V(CO)_6]^-$ is 1858 cm^{-1} and that of $[Mn(CO)_5]^+$ is at 2095 cm^{-1}.

3.3 **Alkene complexes**

1. In general, if ML$_x$ needs n number of electron pairs to obey the EAN rule, it will bond to an n number of alkene or to a portion of a poly-alkene with more than the n number of π bonds.

   The iron carbonyl fragment, Fe(CO)$_3$, has 14 electrons (8 from Fe + 6 from the CO's). Therefore, it needs four more electrons or two pairs to attain its EAN. Consequently, it will bond to dienes so as to obey the EAN rule (Fig. 10.9).

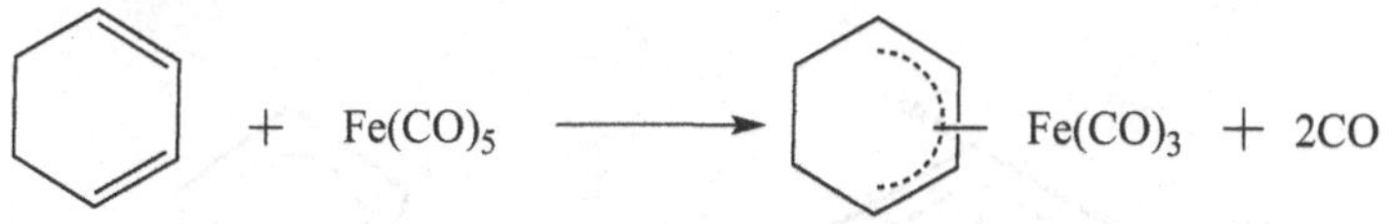

■ **FIGURE 10.9** Formation of tricarbonyl(1-4-η^4-cyclohexadiene)iron(0) complex.

In most cases, π bonds of the olefin must be conjugated to link with the metal moiety. Some examples leading to formation of stable iron and chromium metal carbonyl complexes are shown in Fig. 10.10.

2. If ML_x needs an odd number of electrons to attain its EAN, it will bond with an allyl or a polyenyl group.

1,3,5-cycloheptetriene + Fe(CO)$_5$ → [Fe(CO)$_3$ complex] + 2CO

Fe(CO)$_3$

tricarbonyl(1-4-η^4-1,3,5-cycloheptatriene)iron(0)

+ Cr(CO)$_6$ → [Cr(CO)$_3$ complex] + 3CO

Cr(CO)$_3$

tricarbonyl(η^6-1,3,5-cycloheptatriene)chromium(0)

■ **FIGURE 10.10** Formation of tricarbonyl complexes of iron(0) and chromium(0).

The most common allyl radicals are three-electron linear and five-electron cyclic donors as shown previously in Fig. 10.1 and an example of five-electron linear donor is shown in Fig. 10.11. Examples of cobalt and manganese tricarbonyl complexes are shown in Fig. 10.12 in which the donor ligands shown in Fig. 10.1 have been incorporated.

5 electron "linear" donor

■ **FIGURE 10.11** Five-electron linear donor radicals.

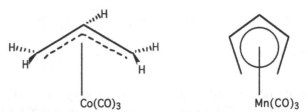

Co(CO)$_3$ Mn(CO)$_3$

■ **FIGURE 10.12** Tricarbonyl complexes of Co and Mn with linear and cyclic donor radicals.

The $Co(CO)_3$ fragment has 15 valence electrons. Therefore it needs three electrons to attain EAN and it forms a stable complex with the allyl radical ($C_3H_5^\bullet$). Conversely, $Mn(CO)_3$ has 13 valence electrons, and it needs 5 more electrons to attain EAN and, therefore, it forms a stable complex with cyclopentadienyl (Fig. 10.12).

3.4 **Aromatic complexes**

1. Cyclic aromatic compounds can act as donors toward transition metals so as to form *sandwich* compounds.
2. The most common aromatic six- and four-electron donors are shown in Fig. 10.13. The corresponding metal sandwich complexes should obey the EAN rule. Ferrocene and chromocene are shown, as examples, within Fig. 10.14.

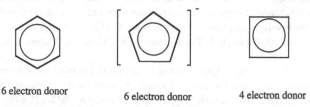

6 electron donor   6 electron donor   4 electron donor

■ **FIGURE 10.13** Six- and four-electron aromatic donor ligands.

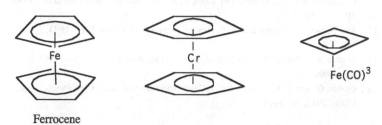

Ferrocene

■ **FIGURE 10.14** Sandwich geometry of *d*-block metal complexes with aromatic donor ligands.

For ferrocene (Fig. 10.14), it is an open question as to whether the molecule is best considered as an iron(0) sandwiched by two cyclopentadienyl radicals or an iron(II) sandwiched by two anionic $C_5H_5^-$ ligands. Ferrocene can be prepared by the following reaction:

$$FeCl_2 + 2NaC_5H_5 \rightarrow \left(\eta^5 - C_5H_5\right)_2Fe + 2NaCl$$

BIBLIOGRAPHY

1. Schaller, C. P.; Grahm, K. J.; Johnson, B. J. Modules for Introducing Organometallic Reactions: A Bridge Between Organic and Inorganic Chemistry. *J. Chem. Educ.* **2015,** 986–992.
2. Parshall, G. W.; Putscher, R. E. Organometallic Chemistry and Catalysis in Industry. *J. Chem. Educ.* **1986,** *63* (3), 189–191.
3. Gilman, H. Organometallic Compounds. *J. Am. Chem. Soc.* **1959,** 1–9.
4. Rochow, E. G. The Direct Synthesis of Organometallic Compounds. *J. Chem. Educ.* **1966,** *43* (2), 58–62.
5. Cotton, F. A. Proposed Nomenclature for Olefin-metal and Other Organometallic Complexes. *J. Am. Chem. Soc.* **1968,** *90* (22), 6230–6232.
6. Fernelius, W. C.; Loening, K.; Adams, R. M. Notes on Nomenclature: Differences between "Organic" and "Inorganic" Nomenclature. *J. Chem. Educ.* **1976,** *53* (12), 773–774.
7. Cioslowski, J.; Hay, P. J.; Ritchie, J. P. Charge Distributions and Effective Atomic Charges in Transition-Metal Complexes Using Generalized Atomic Polar Tensors and Topological Analysis. *J. Phys. Chem.* **1990,** *94,* 148–151.
8. Sun, C. Q.; Sun, Y.; Nie, Y. G.; et al. Coordination-resolved C−C Bond Length and the C 1s Binding Energy of Carbon Allotropes and the Effective Atomic Coordination of the Few-layer Graphene. *J. Phys. Chem.* **2009,** *113,* 16464–16467.
9. Ameen, J. G.; Durfee, H. F. The Structure of Metal Carbonyls. *J. Chem. Educ.* **1971,** *48* (6), 372–375.
10. Blom, R.; Enthaler, S.; Inour, S.; Irran, E.; Driess, M. Electron-Rich N-Heterocyclic Silylene (NHSi)-Iron Complexes: Synthesis, Structures, and Catalytic Ability of an Isolable Hydridosilylene-iron Complex. *J. Am. Chem. Soc.* **2012,** *135,* 6703–6713.
11. Chukwu, R.; Hunter, A. D.; Santarsiero, B. D. Novel Organometallic Complexes Containing Aromatic Azines: Synthesis and X-ray Crystal Structure of 4,6-Bis[(n^5-cyclopentadienyl)dicarbonyliron] 2-(Methylthio)pyrimidine. *Organometallics* **1991,** *10* (7), 2141–2152.
12. Maslowsky, E. Organometallic Benzene Complexes. *J. Chem. Educ.* **1993,** *70* (12), 980–984.
13. Wynne, K. J. An Introduction to Inorganic and Organometallic Polymers. *ACS Symp. Ser.* **1988,** *360,* 1–4.
14. Gasser, G.; Ott, I.; Metzler-Nolte, N. Organometallic Anticancer Compounds. *J. Med. Chem.* **2011,** *54,* 3–25.

Catalysis With Organometallics

1. INTRODUCTION: WHAT IS SO SPECIAL ABOUT CATALYTIC PROCESSES IN OUR DAILY LIFE?

A catalyst is a substance that accelerates the rate of a chemical reaction by lowering the activation energy and helps the reaction to proceed in the forward direction without converting itself into any products, as it can be recovered unchanged at the end of a reaction. Thus, the overall process is known as "catalysis". Many chemical and pharmaceutical industries have adopted catalysis in the production of important chemicals and drugs for our daily usage. Thus, catalysis is highly important in the areas of pharmaceuticals, organometallics, environmental science, materials science, biomaterials, nanotechnology, and in many other disciplines of science and engineering. To create a healthy environment around the globe, scientists have developed catalytic converters for use in automobiles and industrial exhaust systems to reduce the air pollution, to reverse the ozone depletion, etc. Therefore, many *d*-block metal compounds and their organometallic complexes came into existence for use as important catalysts in the areas of heterogeneous catalysis, homogeneous catalysis, organocatalysis, electrochemical catalysis, autocatalysis, nanomaterial-based catalysis, tandem catalysis, enzymatic biocatalysis, photocatalysis, acid/base catalysis, etc. Some monoclonal antibodies, known as "abzymes", have been studied in cancer therapies and enzyme-based biocatalysis has been implemented in the production of high-fructose corn syrup and acrylamide. This chapter focuses on these types of catalysts and their uses.

2. HOMOGENEOUS CATALYSTS

2.1 Oxidative addition and reductive elimination

Most homogeneous catalysis reactions involve two processes known as Oxidative Addition and Reductive Elimination as shown in Fig. 11.1.

Advanced Inorganic Chemistry. http://dx.doi.org/10.1016/B978-0-12-801982-5.00011-4

■ **FIGURE 11.1** Oxidative addition and reductive elimination.

1. In this type of homogeneous catalytic process, XY could be X_2, H_2, HX, RX, etc., L is a ligand, and M is a transition metal that can increase its oxidation number by two and increase its coordination number by two.
   - **a.** The square planar d^8 system (among others) has 16 valence electrons; by accepting two more ligands and "losing" two electrons via oxidation, they can achieve 18 valence electrons.
   - **b.** Oxidative addition is where the oxidation number and coordination number *increases* by two.
   - **c.** Reductive elimination is where the oxidation number and coordination number *decreases* by two.
2. An example of a catalyst for oxidative addition and reductive elimination is the Vaska's complex, a trivial name for a chemical compound discovered by Lauri Vaska, which has the formula IrCl(CO)$[P(C_6H_5)_3]_2$. In the systematic International Union of Pure and Applied Chemistry (IUPAC) nomenclature the name is *trans*-carbonylchloro-bis(triphenyl-phosphine)iridium(I). A generalized example of its use is shown in Fig. 11.2. It is a square planar diamagnetic organometallic complex with iridium metal center and it exists as a bright yellow crystalline solid. In the synthesis, triphenylphosphine ($P(C_6H_5)_3 = L$) serves as both a ligand and a reductant, and the carbonyl ligand is

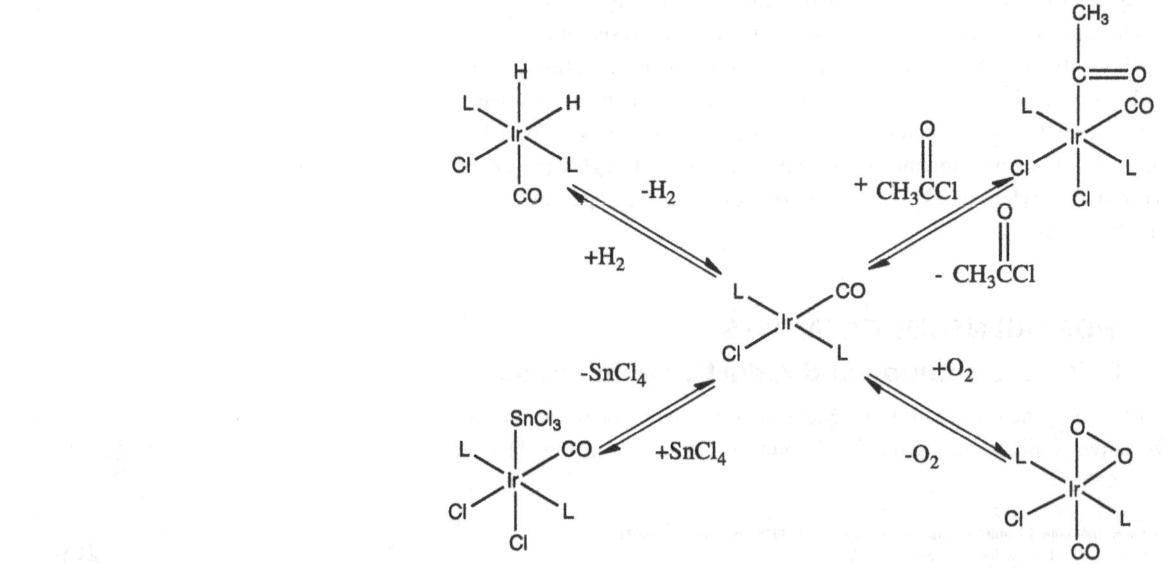

■ **FIGURE 11.2** Oxidative addition and reductive elimination using Vaska's complex.

derived from the decomposition of dimethylformamide ($HCON(Me)_2$), probably via a deinsertion of an intermediate Ir-C(O)H species.

2.2 Insertion reaction (ligand migration reactions)

1. Addition of an organometallic compound, M—X, to an unsaturated molecule, Y, to form a new compound where Y is inserted between M and X is the ligand migration reaction or an example of an insertion reaction as shown in Eq. (11.1).

$$M—X + Y \longrightarrow M—Y—X \qquad (11.1)$$

a. Y = CO, $\overset{}{\underset{}{\text{C}}}$=$\overset{}{\underset{}{\text{C}}}$, diene, alkyne, $R—\overset{\overset{O}{\parallel}}{C}—H$, RCN, SO₂, O₂, or one of many other unsaturated molecules.

b. $X = H^-, R^-, OR^-, NR_2^-, OH^-, H_2O,$ or X^-.

c. The effect of the reaction is that M—X adds across the multiple bond present within Y, as in Eq. (11.1).

2. Showing more detail, the general mechanism involves Y coordinating with the metal M and X, then migrating as shown in Eq. (11.2).

$$(11.2)$$

3. An example of this type of catalytic process is shown in Eq. (11.3).

$$CH_3Mn(CO)_5 + CO \longrightarrow CH_3\overset{\overset{O}{\parallel}}{C}Mn(CO)_5 \qquad (11.3)$$

a. If labeled *CO is used as the attaching CO, one finds that the already present nonradioactive CO is inserted (Eq. 11.4).

$$(11.4)$$

b. Other entering groups, such as amines and phosphines, can also be used to cause the migration of CO.

3. HYDROGENATION CATALYSTS

There are different types of hydrogenation catalysts:

1. Complexes without hydride, such as $RhCl(PPh_3)_3$ $(Ph = C_6H_5)$, that can add H_2 by oxidative addition.
2. Complexes with M—H bonds that do not add H_2, but H_2 is added in a later step in the catalytic process. Examples include $RhH(CO)(PPh_3)_3$ and $RuHCl(PPh_3)_3$.

3.1 Wilkinson's catalyst, RhCl(PPh₃)₃

This is an example of the use of a hydrogenation catalyst for conversion of an olefin to an alkane (Eq. 11.5).

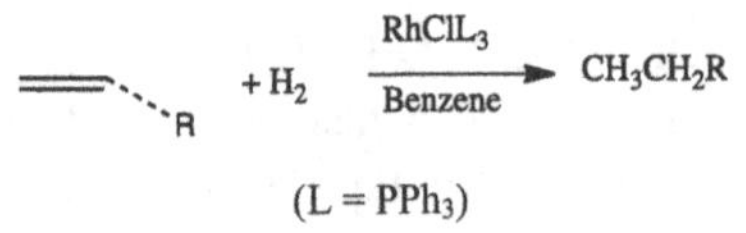

$$\text{olefin} + H_2 \xrightarrow[\text{Benzene}]{\text{RhClL}_3} CH_3CH_2R$$

$$(L = PPh_3)$$

$$(11.5)$$

A schematic mechanism for the hydrogenation is shown in Fig. 11.3.

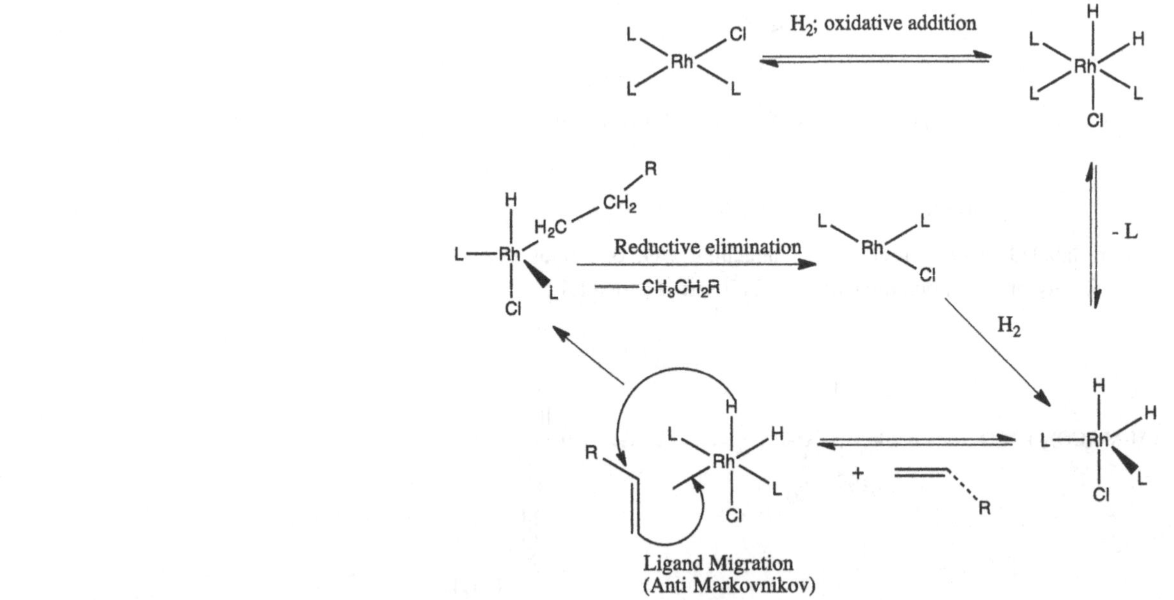

■ FIGURE 11.3 Hydrogenation mechanism with Wilkinson's catalyst.

If solvent molecules (S) are involved, they will form higher coordinated solvated species (solvo-complexes) by complexation with the lower coordinated catalyst as shown in Fig. 11.4.

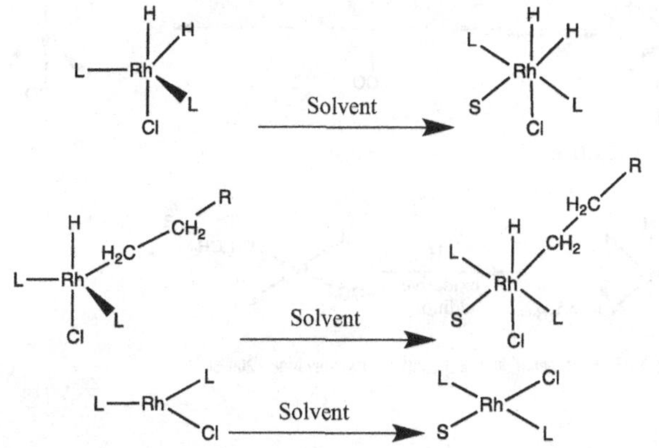

■ **FIGURE 11.4** Formation of higher coordinated solvo-complex.

The important characteristics of the catalytic species are:

1. The high lability of a ligand, which should be able to exist in several forms because of dissociation.
   a. Ph₃P is a bulky ligand and, therefore, its steric bulk helps to bring about early dissociation. If a less sterically hindered phosphine is used, the complex will lose its catalytic activity.
   b. Vaska's complex does not act as a catalyst for this transformation (although it can undergo oxidation addition and reductive elimination with H_2) because of the Ir—PPh₃ bond that is too stable.
2. The ability to exist in several different geometries with coordination number 6 → 5 → 4 or 3.
3. The ability to undergo oxidative addition and reductive elimination.

3.2 Monohydride complexes

Examples include [RhH(CO)(PPh₃)₃], [RuHCl(PPh₃)₃], [IrH(CO)(PPh₃)₃], and [HCo(CO₄)] (from H_2 + Co₂(CO)₈). The rhodium monohydrido complexes can catalyze hydrogenation through oxidative addition and reductive elimination processes as shown in Fig. 11.5.

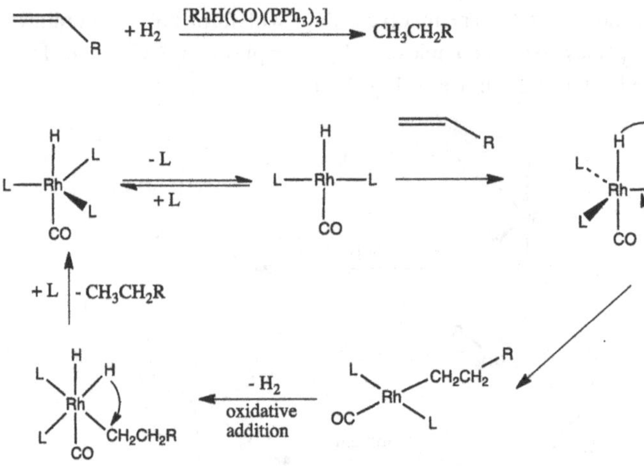

■ **FIGURE 11.5** Hydrogenation using rhodium monohydride catalyst.

3.3 **Hydroformylation and other closely related oxo processes**

1. General requirements are: Co salts, H_2, CO, and organic substrates.
   a. Hydroformylation

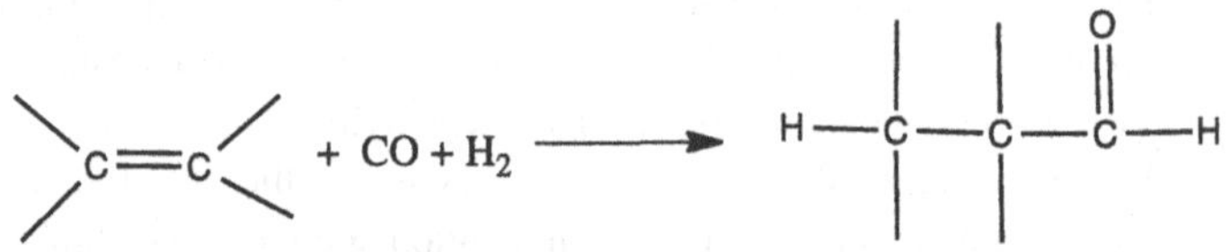

$$\text{(11.6)}$$

2. Other closely related oxo-processes are homologation (Eq. 11.7), hydrogenolysis (Eq. 11.8), and hydrogenation (Eq. 11.9).
   a. Homologation

$$—CH_2OH + CO + 2H_2 \longrightarrow —CH_2CH_2OH + H_2O \quad (11.7)$$

   b. Hydrogenolysis

$$R_2CHOH + H_2 \longrightarrow R_2CH_2 + H_2O \quad (11.8)$$

c. Hydrogenation

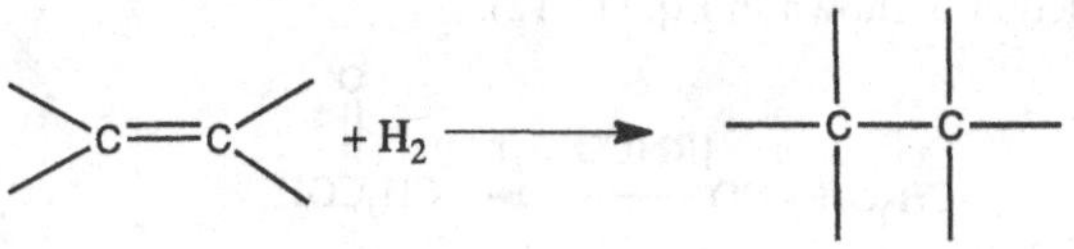

(11.9)

Hydroformylation using a cobalt carbonyl compound is shown in Eq. (11.10).

$$H_2C{=}CH_2 + CO + H_2 \xrightarrow[\substack{90\text{–}200^\circ \\ 100\text{–}400\ \text{atm}}]{Co_2(CO)_8} CH_3CH_2\overset{\overset{\displaystyle O}{\|}}{C}H \qquad (11.10)$$

In the chemistry of Eq. (11.10), first the dicobalt octacarbonyl (also known as cobalt carbonyl), $Co_2(CO)_8$, will react with hydrogen to form a catalytically active compound, $HCo(CO)_4$, as in Eq. (11.11), which will undergo hydroformylation with ethene to form the desired aldehyde as shown in Fig. 11.6.

$$Co_2(CO)_8 + H_2 \longrightarrow 2HCo(CO)_4 \qquad (11.11)$$

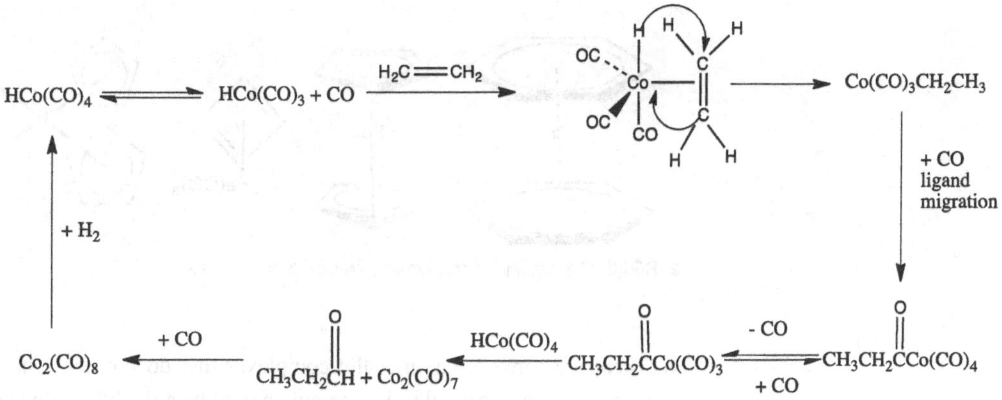

■ **FIGURE 11.6** Hydroformylation of ethene using dicobalt octacarbonyl catalyst.

4. OTHER CATALYTIC PROCESSES

4.1 **Production of acetic acid from CH₃OH**

1. The reaction is shown in Eq. (11.12).

$$CH_3OH + CO \xrightarrow{[Rh(CO)_2I_2]^-} CH_3\overset{\displaystyle O}{\overset{\displaystyle \|}{C}}OH \tag{11.12}$$

2. The proposed reaction pathway is shown in Fig. 11.7. The first step involves the addition of HI to methanol to produce the reactive CH₃I.

$$CH_3OH + HI \longrightarrow CH_3I + H_2O$$

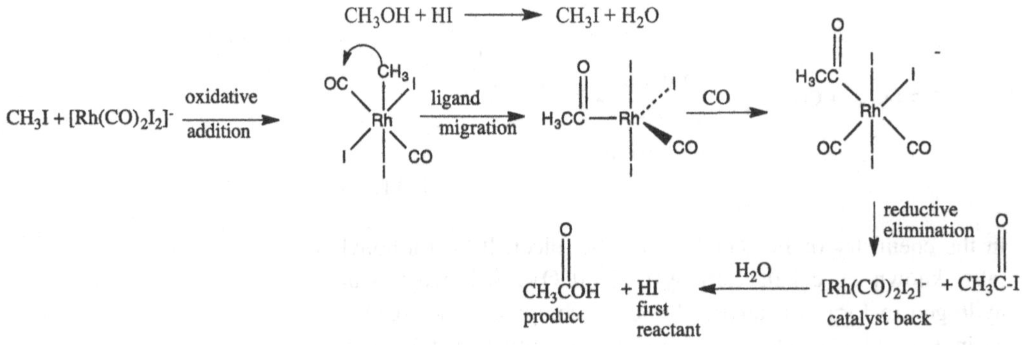

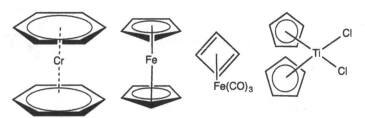

■ **FIGURE 11.7** Formation of acetic acid from methanol using a rhodium catalyst.

3. Cyclohexatriene, cyclopentadiene, and cyclobutadiene sandwiched organometallic compounds (Fig. 11.8) have also been used as catalysts for a number of organic transformations.

■ **FIGURE 11.8** Sandwiched organometallic compounds of Cr, Fe, and Ti.

4. There are some organometallic complexes that do not obey the effective atomic number rule. An example would be a diallyl-sandwiched nickel(II) complex of 16 electrons, whose structure and synthesis is shown in Eq. (11.13).

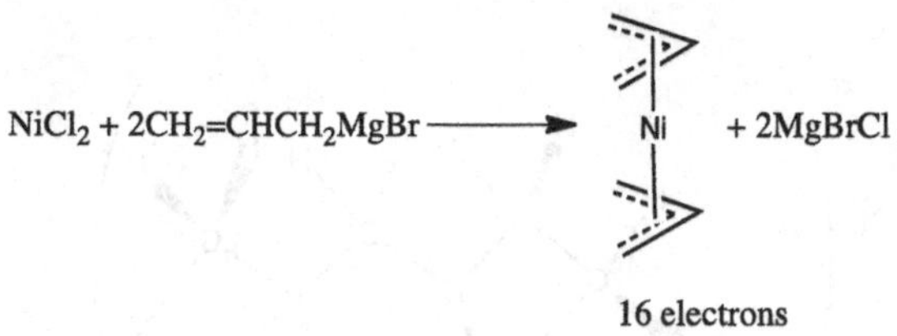
NiCl2 + 2CH2=CHCH2MgBr
Ni
+ 2MgBrCl
16 electrons

(11.13)

4.2 Heterogeneous catalysis of alkene polymerization with Ziegler–Natta catalyst

1. In this catalytic system, a partially alkylated fibrous material, produced from reaction of $TiCl_4$ and $(CH_3CH_2)_3Al$, acts as the catalyst (Fig. 11.9).
2. An important facet in this process is that some of the surface of the Ti metal is not fully coordinated. These sites can serve as a binding target for an alkene, leading to its polymerization.

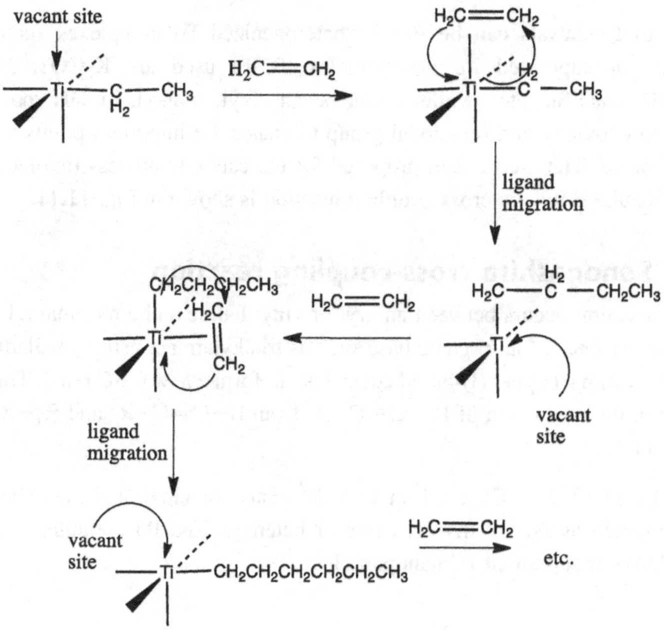
vacant site
Ti
H2C═CH2
ligand migration
vacant site
CH2CH2CH2CH3
Ti—CH2CH2CH2CH2CH2CH3
etc.

■ **FIGURE 11.9** Heterogeneous catalysis using the Ziegler–Natta catalyst.

3. This catalytic process produces very stereoregular polymers when un-symmetrical alkenes are used. Thus, the product of polymerization of $CH_2{=}CHCH_3$ is a stereoregular polymer, as shown in Fig. 11.10.

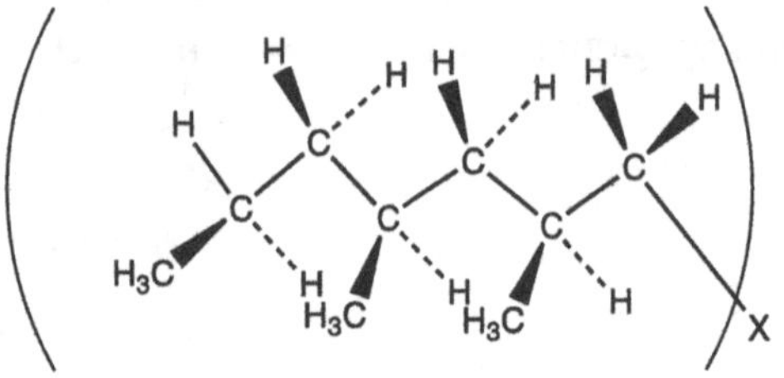

■ **FIGURE 11.10** Repeating unit of a stereoregular polymer from 1-propene.

4.3 Suzuki—Miyaura cross-coupling reaction

The reaction occurs between a boronic acid and an organohalide, and is catalyzed by a palladium-based catalyst in the presence of a base to form a new C—C bond as shown in Eq. (11.14) for production of R_1—R_2.

$$R_1BY_2 + R_2X \xrightarrow{\text{Catalyst}} R_1{-}R_2 + XBY_2 \qquad (11.14)$$

Pd-based catalysts can be free or heterogenized Pd complexes such as $PdAr_3$ or supported Pd nanoparticles. Bases used are K_2CO3, NaF, NaOH, NaO^{t}Bu, etc., while Y can be an alkyl, o-alkyl, or OH moiety. The low toxicity and functional group tolerance are important points to be considered. The mechanism proposed for the catalytic process involved in the Suzuki—Miyaura cross-coupling reaction is shown in Fig. 11.11.

4.4 Sonogashira cross-coupling reaction

This reaction occurs between an aryl or vinyl halide and a terminal alkyne in the presence of an organic base such as trialkylamine, using a palladium catalyst with a copper (I)-based cocatalyst, to form a new C—C bond. This is shown, for production of $R^1{-}C{\equiv}C{-}R^2$ from $H{-}C{\equiv}C{-}R^2$ and $R_1{-}X$, in Fig. 11.12.

In Fig. 11.12, X = Cl, Br, I, or OTf, R^2 = aryl or vinyl, and the catalyst, represented as Pd^0L_2, involves free or heterogenized Pd complexes such as $PdAr_3$ or supported Pd nanoparticles.

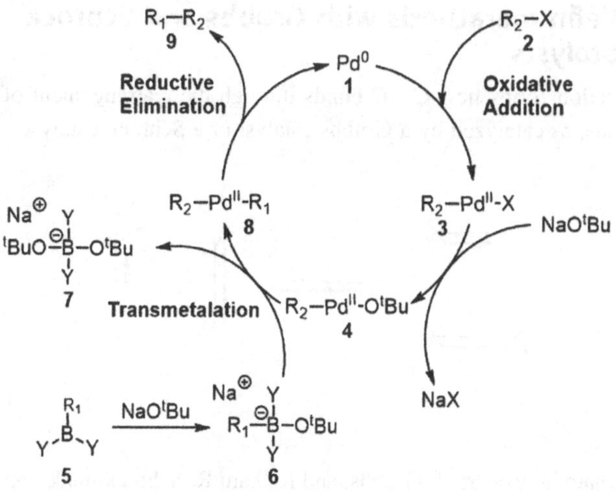

■ **FIGURE 11.11** The catalytic cycle for the Suzuki–Miyaura cross-coupling.

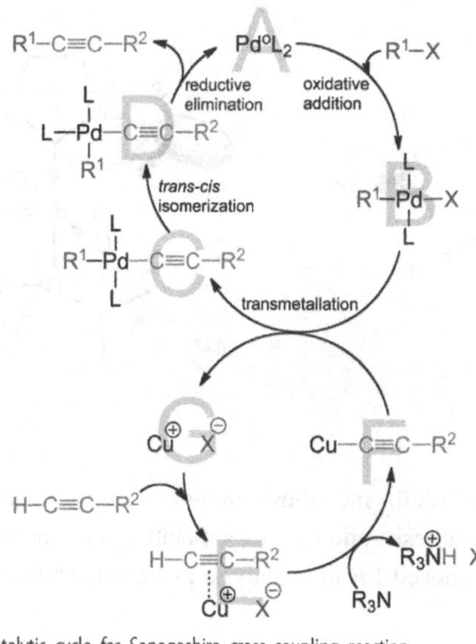

■ **FIGURE 11.12** Catalytic cycle for Sonogashira cross-coupling reaction.

4.5 **Olefin metathesis with Grubbs and Schrock catalysts**

The reaction forms new C=C bonds through the rearrangement of olefin fragments, as catalyzed by a Grubbs catalyst or a Schrock catalyst.

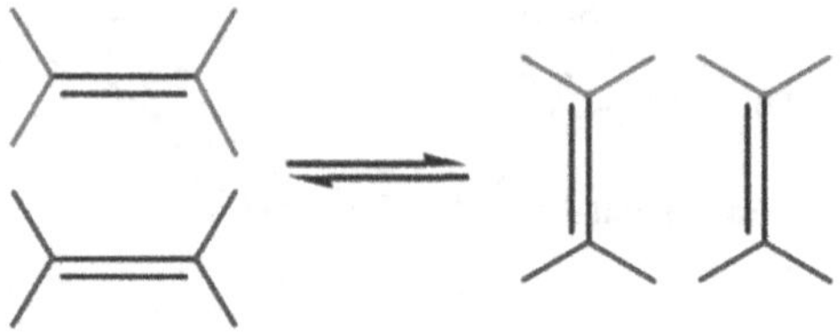

$$(11.15)$$

Yves Chauvin, Robert H. Grubbs, and Richard R. Schrock made significant contributions in the metathesis reaction regarding elucidation of mechanism and catalyst development, and thus they were collectively rewarded the 2005 Nobel Prize in Chemistry.

Typical Grubbs catalysts and Schrock catalysts for metathesis processes are shown in Figs. 11.13 and 11.14.

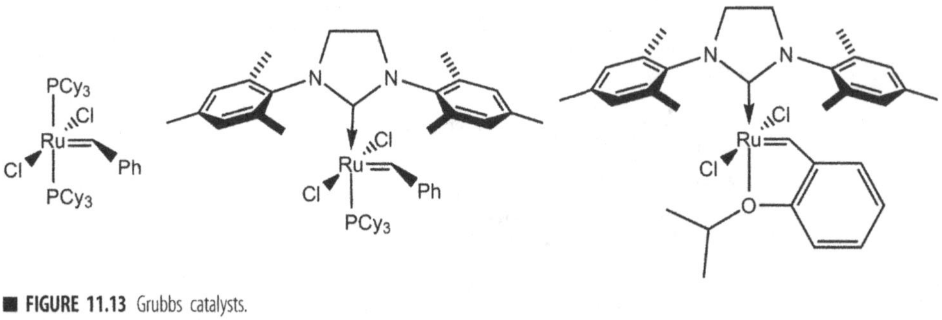

■ **FIGURE 11.13** Grubbs catalysts.

Important classes of olefin metathesis include (1) ring-opening metathesis, (2) ring-closing metathesis, and (3) cross metathesis as shown in Fig. 11.15. The mechanism proposed for the catalytic process is shown in Fig. 11.16.

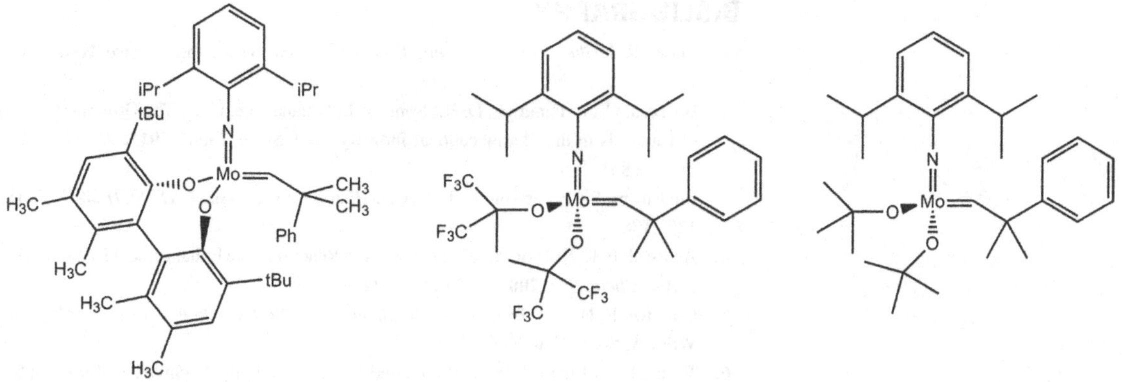

■ **FIGURE 11.14** Schrock catalysts.

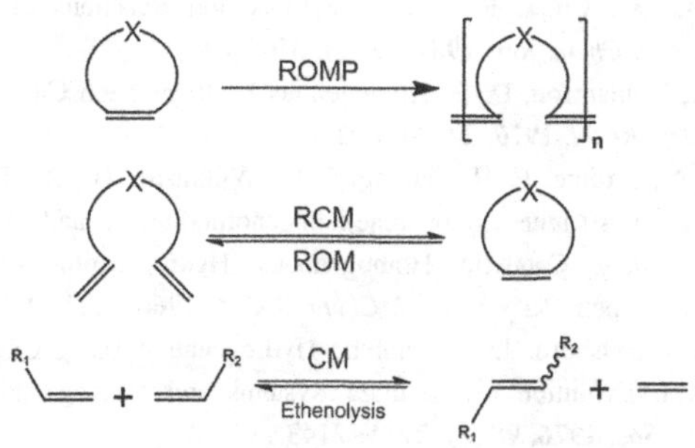

■ **FIGURE 11.15** Three important classes of olefin metathesis.

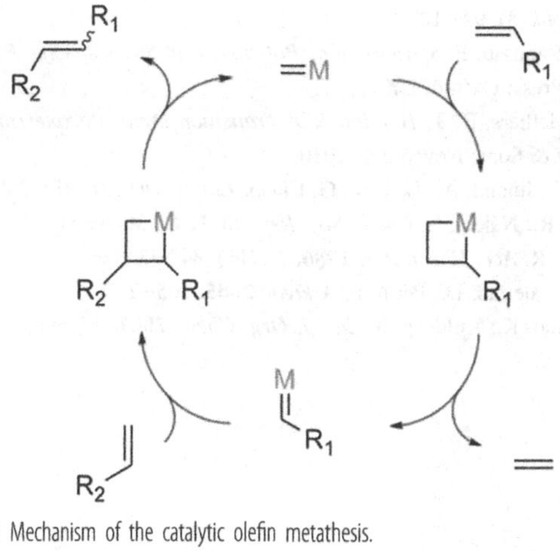

■ **FIGURE 11.16** Mechanism of the catalytic olefin metathesis.

BIBLIOGRAPHY

1. Masel, R. I. *Chemical Kinetics and Catalysis,* 1st ed.; Wiley-Interscience: New York, 2001.
2. Busacca, C. A.; Fandrick, D. R.; Song, J. J.; Senanayake, C. H. The Growing Impact of Catalysis in the Pharmaceutical Industry. *Adv. Synth. Catal.* **2011,** *353* (11–12), 1825–1864.
3. Lindstrom, B.; Pettersson, L. J. A Brief History of Catalysis. *CATTECH* **2003,** *7* (4), 130–138.
4. Agarwal, P. K. Role of Protein Dynamics in Reaction Rate Enhancement by Enzymes. *J. Am. Chem. Soc.* **2005,** *127* (43), 15248–15256.
5. Crabtree, R. H. *The Organometallic Chemistry of the Transition Metals,* 5th ed.; John Wiley & Sons: New York, 2009.
6. Vaska, L.; DiLuzio, J. W. Carbonyl and Hyrdido-Carbonyl Complexes of Iridium by Reaction With Alcohols. Hydrido Complexes by Reaction with Acid. *J. Am. Chem. Soc.* **1961,** *83* (12), 2784–2785.
7. Hartwig, J. F. *Organotransition Metal Chemistry, from Bonding to Catalysis;* University Science Books: New York, 2010.
8. Anderson, G. K.; Cross, R. J. Carbonyl-insertion Reactions of Square Planar Complexes. *Acc. Chem. Res.* **1984,** *17* (2), 67–74.
9. Birch, A. J.; Williamson, D. H. Homogenous Hydrogenation Catalysts in Organic Synthesis. *Org. React.* **1976,** *24* (1), 1–186.
10. Osborn, J. A.; Jardine, F. H.; Young, J. F.; Wilkinson, G. The Preparation and Properties of Tris(triphenylphosphine)halogenorhodium(I) and Some Reactions Thereof Including Catalytic Homogeneous Hydrogenation of Olefins and Acetylenes and Their Derivatives. *J. Chem. Soc. A.* **1966,** 1711–1732.
11. Schrock, R. R.; Osborn, J. A. Catalytic Hydrogenation Using Cationic Rhodium Complexes. I. Evolution of Catalytic Systems and Hydrogenation of Olefins. *J. Am. Chem. Soc.* **1976,** *98* (8), 2134–2143.
12. Ojima, I.; Tsai, C.-Y.; Tzamarioudaki, M.; Bonafoux, D. The Hydroformylation Reaction. *Org. React.* **2000,** *56* (1), 1–354.
13. Jones, J. H. The Cativa Process for the Manufacture of Acetic Acid. *Platinum Met. Rev.* **2000,** *44* (3), 94–105.
14. Natta, G.; Danusso, F. *Stereoregular Polymers and Stereospecific Polymerizations;* Pergamon Press: Oxford, UK, 1967.
15. Hoff, R.; Mathers, R. T. *Handbook of Transition Metal Polymerization Catalysts;* John Wiley & Sons: New York, 2010.
16. Amatore, C.; Jutand, A.; Le Duc, G. *Chem. Eur. J.* **2011,** *17* (8), 2492–2503.
17. Chinchilla, R.; Nájera, C. *Chem. Soc. Rev.* **2011,** *40,* 5084–5121.
18. Schrock, R. R. *Acc. Chem. Res.* **1986,** *19* (11), 342–348.
19. Ileana, D.; Valerian, D.; Petru, F. *Arkivoc* **2005,** 105–129.
20. Sambasivarao, K.; Kuldeep, S. *Eur. J. Org. Chem.* **2007,** *35,* 5909–5916.

Advanced Topics-4:
Bioinorganic Chemistry
and Applications

12

Bioinorganic Chemistry and Applications

1. INTRODUCTION

Bioinorganic chemistry is a field that encompasses the intersection between inorganic chemistry and biochemistry. Inorganic molecules, including metal ions and coordination compounds, are necessary for life in many organisms, as they function in the transport of molecules and are a key component of enzymes. Inorganic molecules have also been used in compounds that have successfully treated cancer, pernicious anemia, and Alzheimer disease. Despite the importance of these molecules, transition metals, including cobalt, copper, nickel, molybdenum, and chromium, are found in the human body in only very small amounts. Of the transition metals found in human physiology, iron is the most abundant, adding up to as much as 5 g in the body of a healthy adult. Iron is of significant importance as it allows the transportation of oxygen throughout the body, and is stored within molecules for future use. Iron, copper, and zinc are found in all organisms, with few exceptions, whereas other transition metals are only found in specific organisms. A large quantity of compounds within the body such as metalloenzymes, metalloproteins, coenzymes or vitamins, nucleic acids, and hormones also contain inorganic elements.

The following exploration of bioinorganic chemistry will include an overview of its history, an explanation of its incorporation into physiological processes in the human body, its medical applications, and future uses. Although this chapter is not comprehensive of the processes and uses of the principles of inorganic chemistry within the body, it will provide a brief explanation of a few important topics.

2. HISTORY AND MEDICAL RELEVANCE

Although bioinorganic chemistry had been discovered and studied many years previously, with proven use of metals in pharmaceutical potions

Advanced Inorganic Chemistry. http://dx.doi.org/10.1016/B978-0-12-801982-5.00012-6

dating back to the ancient civilizations of Mesopotamia, India, China, and Egypt, the first practical application of this discipline of chemistry was found in agriculture with elements such as phosphorous and nitrogen. Bioinorganic chemistry has since been highly researched and proved to be invaluable for the development of new treatments for diseases, as well as in understanding concepts such as biomineralization and aspects of environmental chemistry.

2.1 **Salvarsan**

The advent of modern medicinal chemistry is typically attributed to Paul Ehrlich's discovery of the organometallic compound salvarsan, a strong antibiotic that was originally used in the treatment of syphilis. Salvarsan was first synthesized in 1907 as part of a systematic effort by Ehrlich's laboratory to screen hundreds of compounds in the search of *magic bullets*: drugs that exhibit antimicrobial activity without affecting the human patient. In 1912, Ehrlich and his colleagues published the first results showing the antisyphillis properties of salvarsan. After years of further research and chemical adjustments to salvarsan to improve its pharmaceutical applications, such as the addition of mercury and bismuth, the organoarsenic compounds were eventually replaced by penicillins for the treatment of syphilis after the Second World War. Despite the extensive use and research of salvarsan, its exact chemical composition is still unknown. For years, it had been thought that an As=As double bond occupied the center of the molecule conjugated with two aminophenol moieties, as shown in Fig. 12.1a. However, in 2005, mass spectral analysis showed that the core of salvarsan may in fact be occupied by As—As single bonds, as in the proposed structures for (b) and (c) in Fig. 12.1.

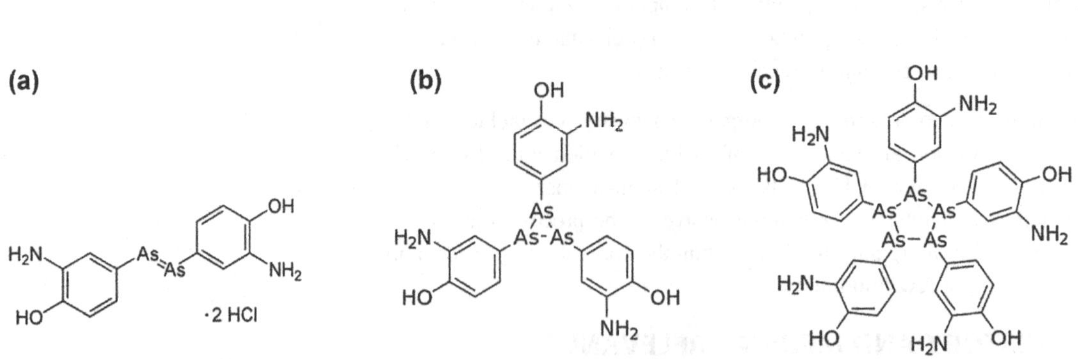

■ FIGURE 12.1 Traditional molecular structure of salvarsan (a), as well as the trimer (b) and pentamer (c) structures suggested by mass spectral analysis.

2.2 **Vitamin B12**

In the late 1920s, Minot and Murphy discovered a successful treatment for the previously fatal disease pernicious anemia. It was found that by supplementing the diet with liver, pernicious anemia could be reversed. The reason for this was because the liver contains cyanocobalamin, or vitamin B12. However, it was not until 1948 that Folkers and Smith were successful in isolating vitamin B12. This molecule, shown in Fig. 12.2, has the most complex monomeric structure seen in nature and has been extensively studied due to its magnitude of uses within the body. It has been determined that the covalent C-Co bonds within the molecule promote catalytic activity. B12 is active as a cofactor for many enzymes, which are involved in reactions relating to nucleic acid and lipid synthesis.

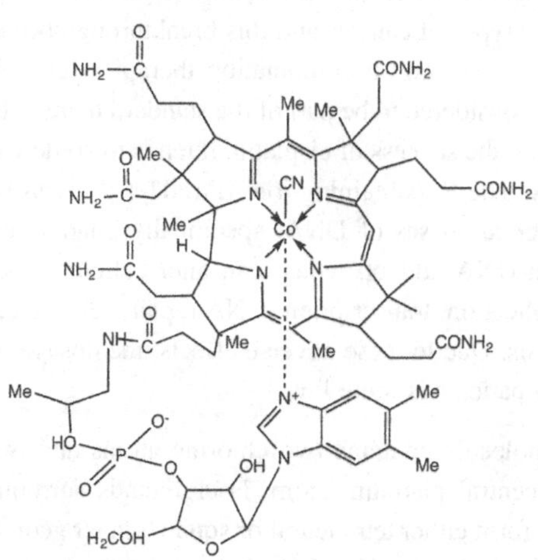

■ **FIGURE 12.2** The structure of vitamin B12. Replacement of the cyano with a methyl group would result in methylcobalamin, a derivative of vitamin B12.

Vitamin B12 belongs to a group of molecules called corrinoids. These molecules contain a central corrin molecule, which is a cyclic system with four pyrrole rings, and has a structure similar to that of heme and other porphyrins. The cobalt atom lies within the plane of the corrin ring, bonded by four ligands. Different substituents on the corrin ring in the beta position result in different molecules. For example, when a CN^- group is attached to this position, it is known as common vitamin B12. This molecule is then a diamagnetic Co (III) system with a d6 configuration. However, when a methyl group is ligated in the beta position, the molecule becomes the biological cofactor methylcobalamin. These compounds related to vitamin B12 are called cobalamins, many of which are important cofactors for biological reactions.

2.3 **Cisplatin and cancer treatments**

One of the first major developments in cancer treatment began with an accidental discovery in 1965 by Dr. Barnett Rosenberg. His research on dividing cells led him to perform further experiments using platinum electrodes and an electric field to study methods to halt cell division. After 2 years of such experiments, he and his colleagues discovered that it was not the electric field that had caused cell division to stop, but the platinum compound being released from the electrodes. This revelation led to the development of cisplatin, a drug used to treat cancerous tumors. The first clinical trials were conducted in 1972, and despite critics' skepticism about using a poisonous heavy metal within a drug, cisplatin was approved for the treatment of testicular and bladder cancers by the US Food and Drug Administration (FDA) in 1978. The uses of cisplatin expanded to include treatment of multiple other types of cancer, and this breakthrough became the start of large amounts of research on combination therapies for cancer treatment. Cisplatin is still considered to be part of the standard therapy for many forms of cancer. Despite the success of cisplatin, it tends to come with many negative side effects. The Pt containing ion, $[Pt(NH_3)_2]^{2+}$, can bind covalently with the nitrogenous bases of DNA, specifically guanine and adenine, to form a cisplatin-DNA adduct, which can alter cellular pathways to result in abnormal replication, transcription, DNA repair, cell cycle, and cell death through apoptosis. Due to these adverse effects, the dosage of cisplatin that can be given to patients is quite low.

The cisplatin molecule contains two chlorine atoms and two NH_3 groups surrounding a central platinum atom. Four ligands surrounding a central metal atom can form either tetrahedral or square planar geometry. However, due to the eight d electrons in Pt^{2+}, it is energetically favorable for the ligands to arrange into a square planar geometry (Fig. 12.3).

■ **FIGURE 12.3** The square planar structure of cisplatin.

Due to the geometry, it is possible to form both *cis-* and *trans-* isomers. To form cisplatin, the chlorine atoms and NH_3 groups must be arranged in a *cis*-orientation, which cannot be achieved by simply adding two NH_3 molecules to a $PtCl_4$ molecule. The *trans-* effect causes the second NH_3 group to orient itself *trans-* to the first NH_3 group, because the Cl atoms have a larger *trans-* effect than do the NH_3 ligands. This effect is utilized to synthesize cisplatin in a reaction pathway such as the one shown in Scheme 12.1.

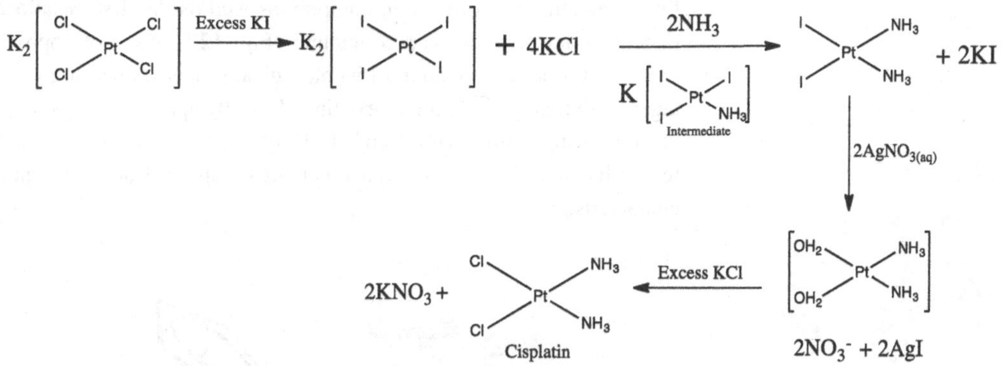

■ **SCHEME 12.1** Scheme for the formation of cisplatin.

Further research of anticancer platinum drugs sought to alleviate the toxicity of cisplatin while maintaining a similar therapeutic effect. Many of these drugs, some shown in Fig. 12.4, were designed to increase hydrolytic stability through the replacement of the two chlorine ligands with carboxylate rings, which minimizes the chemical reactivity as well as the possibility of side effects. Carboplatin, *cis*-[Pt (1,1-dicarboxycyclobutane)$(NH_3)_2$], and oxaliplatin [Pt(1R,2R-1,2-diaminocyclohexane)(oxalate)] have both been approved worldwide, whereas heptaplatin, lobaplatin, and nedaplatin are currently undergoing clinical trials in the United States and have been approved for use in Japan, China, and South Korea.

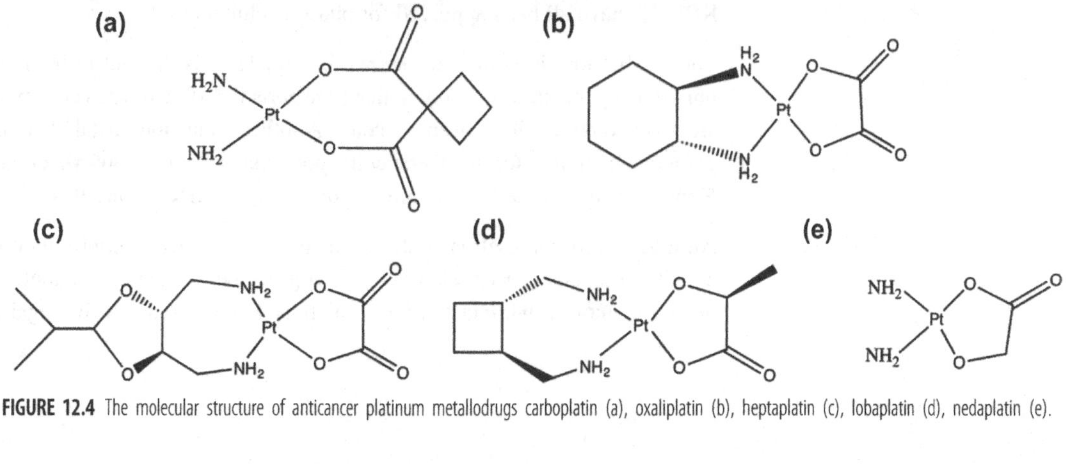

■ **FIGURE 12.4** The molecular structure of anticancer platinum metallodrugs carboplatin (a), oxaliplatin (b), heptaplatin (c), lobaplatin (d), nedaplatin (e).

Despite the adverse effects of cisplatin, the discovery of such a potent and versatile anticancer drug encouraged further exploration into the medicinal applications of other organometallic compounds. Within years, the anticancer activity of transition-metal cyclopentadienyl complexes was investigated by Köpf and Köpf-Maier. Titanocene chloride, $Ti(\eta^5--C_5H_5)_2Cl_2$, emerged from this study as a promising anticancer compound and was the first nonplatinum coordination complex as well as the first metallocene to undergo clinical trials. The compound (Fig. 12.5) was not approved for use after further examination of its biological properties because no advantages were found over those of existing clinically approved drugs. However, the promising results with $Ti(\eta^5-C_5H_5)_2Cl_2$ greatly influenced the future research concerning similar organometallic compounds and their anticancer characteristics.

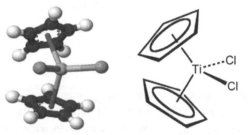

■ **FIGURE 12.5** The molecular structure of titanocene chloride, $Ti(\eta^5-C_5H_5)_2Cl_2$.

Research that followed examined a variety of organometallic compounds, each exhibiting unique anticancer properties. Since 1990, ruthenium complexes in particular have been at the center of anticancer metallodrug research. The ruthenium-based anticancer agents NAMI-A, *trans*-tetrachloro(dimethylsulfoxide)imidazole-ruthenate(III), and KP1019, *trans*-tetrachlorobis(1H-indazole)ruthenate(III), as well as its sodium salt (N) KP1339, have all been approved for phase II clinical trials.

The Ru(III) ion, in complexes shown in Fig. 12.6, is thought to be transported using transferrin, a protein that transports Fe(III) and whose receptors are overexpressed in cancerous cells. Another transition metal that has garnered attention for its therapeutic potential is iron. Köpf-Maier and Köpf investigated the ferrocenium cation $[Fe(\eta^5-C_5H_5)_2]^+$ in 1984.

Although ferrocene itself is nontoxic, the cation has shown antiproliferative activity in several cancer cell lines. This approach of conjugating a nontoxic ferrocene moiety with tamoxifen, which is known to be active against

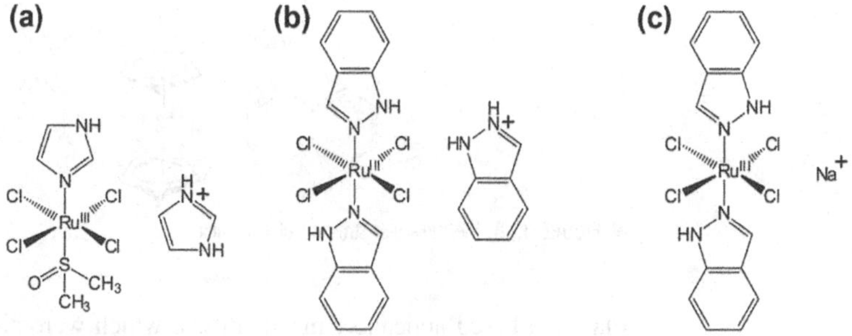

FIGURE 12.6 The molecular structures of NAMI-A (a), K1019 (b), and (N)KP1339 (c).

hormone-dependent breast cancer, and introducing this ferrocifen derivative (Fig. 12.7) into tumor cells, where it is then oxidized to the harmful ionic form, has been further used in the treatment of ER(+) MCF-7 and ER(−) MDA-MB-231 breast cancers.

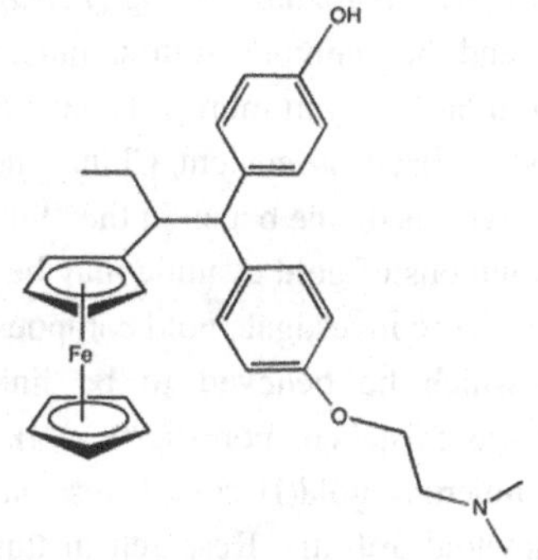

FIGURE 12.7 The molecular structure of ferrocifen.

2.4 Other therapeutic applications of organometallic compounds

Derivatives of ferrocene have also shown therapeutic activity in the treatment of other diseases. The antimalarial drug ferroquine is of the same vein as ferrocifen, this time combining ferrocene with the known antimalarial drug chloroquine. This drug, shown in Fig. 12.8, is currently in phase II clinical trials and it is believed to act through a binary mechanism involving an interaction with free heme with the production of reactive oxygen species.

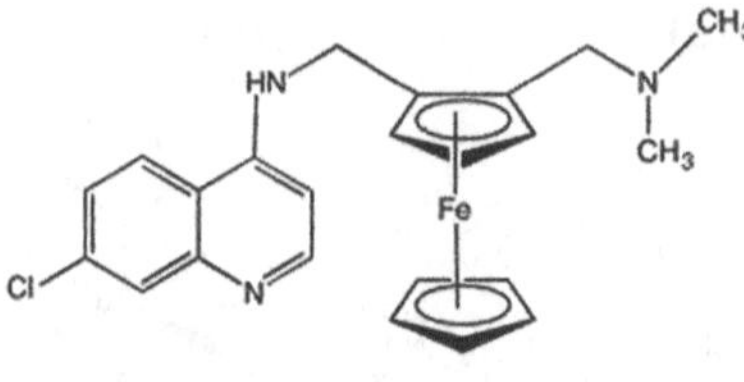

■ **FIGURE 12.8** The molecular structure of ferroquine.

Platinum-based anticancer metallodrugs, which were discussed earlier, are not the only complexes of precious metals that have been used for therapeutic purposes: both gold and silver have a rich history of medicinal applications. Silver has been applied in wound treatment since the 18th century. The antimicrobial activity of silver ions was recognized during this time, and silver nitrate was used in the treatment of ulcers. The use of silver as an antimicrobial agent was approved by the FDA in 1920s; however, its application began to diminish with the introduction of antibiotics in the following decades. Topical silver gained popularity once again in 1968 with the introduction of silver sulfadiazine (Fig. 12.9a), a product of the reaction of silver nitrate and the antibiotic sulfonamide, which is used in the treatment of burns. Gold has an even more extensive medical history, with applications dating as far back as ancient China and India. The use of gold compounds in modern medicine began in the 19th century with Robert Koch's findings that solutions of gold cyanide may be tuberculostatic. This prompted Jacques Forestier to investigate gold compounds as a treatment for rheumatoid arthritis, which he believed to be linked to tuberculosis. Although this theory was mistaken, Forestier's work in the 1930s led to the development of numerous gold(I) complexes that have been used in the treatment of rheumatoid arthritis. Research in this area culminated in the development of auranofin (Fig. 12.9b), (triethylphosphine gold(I) tetraacetyl-thioglucose), an orally administered drug that was approved by the FDA in 1985 and has since seen widespread application as an antirheumatic agent and is currently undergoing clinical evaluation for the treatment of human immunodeficiency virus infection, chronic lymphatic leukemia, and squamous cell lung cancer.

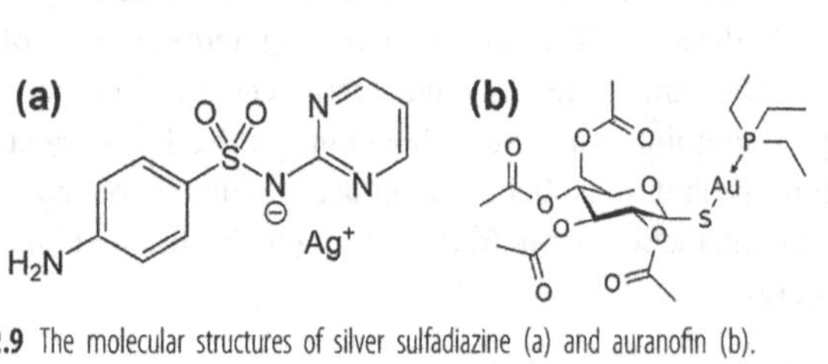

■ **FIGURE 12.9** The molecular structures of silver sulfadiazine (a) and auranofin (b).

2.5 Diagnostic metallodrugs

Metallodrugs have not just been used for therapeutic applications in modern medicine, but also been applied extensively in diagnostic imaging and radio-pharmaceuticals. One of the most powerful techniques used in diagnostic medicine is magnetic resonance imaging (MRI), a noninvasive scanning technique that employs strong magnetic fields and radio waves to obtain three-dimensional images of the human body's anatomic structures, as well as the functional and physiological properties of their tissues. MRI relies on the same concepts as nuclear magnetic resonance by using an external magnetic field to allow the hydrogen nuclei of water in bodily tissues to be polarized according to the magnetic field. By altering the parameters of the applied magnetic field through *gradient pulses*, each distinct tissue can be observed due to differences in the relaxation properties of the hydrogen atoms that they contain. This distinction depends on several factors, namely, proton density, longitudinal reaction time ($1/T_1$), and transverse relaxation time ($1/T_2$). Since the invention of MRI in 1973, many MRI contrast agents have been developed to improve image resolution by catalytically shortening the relaxation time of the water molecules in the targeted area.

The most common contrast agents used in MRI are intravenously administered gadolinium organometallic complexes. The 3^+ oxidation state of gadolinium, Gd(III), is highly paramagnetic with seven unpaired electrons ($4f^7$) and it has a long electronic relaxation time. These characteristics allow Gd(III) to act as a relaxation agent, effectively reducing the longitudinal relaxation times of the protons in water, which results in an enhanced signal and image brightening. These *positive* gadolinium (III) chelate contrast agents are administered in approximately one-third of all MRI scans, namely, Gd(DTPA)(H$_2$O)]$^{2-}$, Gd^{3+}-diethylenetriaminepentaacetic, and [Gd(DOTA)(H$_2$O)]$^-$, Gd^{3+}-tetraazacyclododecanetetraacetic acid, two of which are shown in Fig. 12.10.

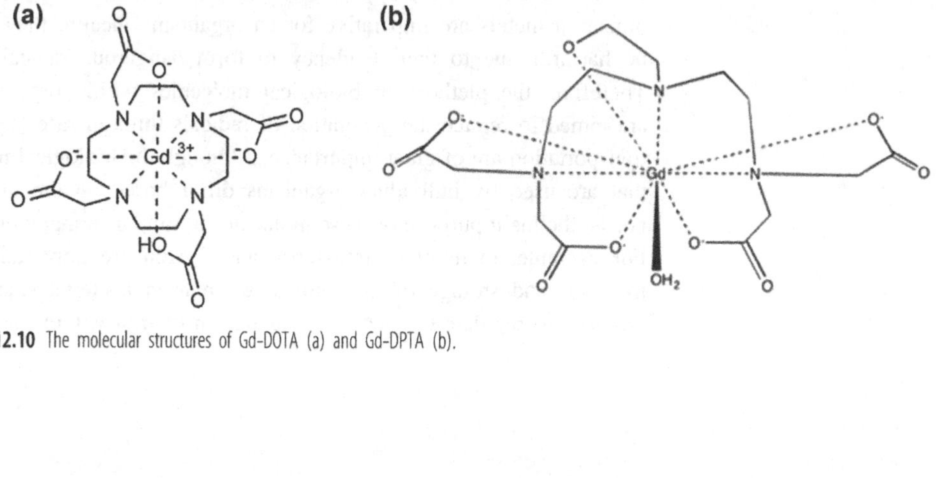

■ **FIGURE 12.10** The molecular structures of Gd-DOTA (a) and Gd-DPTA (b).

Another class of MRI contrast agents decreases transverse relaxation time and this results in image darkening. These *negative* MRI contrast agents largely use superparamagnetic iron oxide nanoparticles. Although there are many iron oxide contrast agents that have been approved in the past, the only agent still approved for use is the Gastromark. This orally administered contrast agent was approved by the FDA in 1996.

Radiopharmaceuticals are another class of metallodrug that play an important role in medical diagnosis. These radioactive agents have become a powerful tool used in the diagnosis of cancer, cardiological and neurological disorders, as well as kidney and liver abnormalities. To image such a variety of conditions, a number of different imaging agents are used to target a specific organ or body fluid. The most widely used radioisotope in diagnostic imaging is technetium-99m, which is used in 80% of all nuclear medicine procedures. The 6-h half-life of ^{99m}Tc is ideal for medical application as it is long enough to evaluate metabolic processes while being short enough to minimize the adverse effects of radiation on the patient. The decay of ^{99m}Tc is also isomeric, emitting 140 keV γ-rays, which are then detected by a gamma camera for diagnostic imaging, as well as low energy electrons. The absence of high-energy beta-emission again minimizes the effects of radiation on the patient. Another advantage of technetium is its versatile chemistry and variety of oxidation states. The nine oxidation states of technetium and flexible stereochemistry allow the formation of complexes with coordination numbers between four and nine, allowing the coordination of a number of biologically active compounds to form target-specific tracers that seek out a specific tissue or organ.

3.   TRANSPORT AND STORAGE OF METAL IONS

The field of bioinorganic chemistry also encompasses the mechanisms by which metal ions are transported and stored within the body. Although transition metals are imperative for an organism's health, they can also be harmful due to their tendency to form dangerous radical species. Therefore, the plethora of biological molecules within organisms that are aimed to reduce the generation of radicals through safe storage and transportation are of great importance. Although the biological molecules that are used by individual organisms differ based on the organism's needs, the main purpose of these molecules is similar in many organisms. For example, in humans, transferrin and ferritin are important for the transport and storage of iron atoms, whereas in bacteria, siderophores are used to regulate the actual concentration of iron within the bacteria.

In addition, metal ions play an important role in the signaling mechanisms throughout the body. This involves the influx and efflux of Na^+ and K^+ ions, a process that is essential for our overall health and functioning.

3.1 **Iron storage: transferrin**

Iron is transported throughout the body in small glycoproteins called transferrins. There are various types of transferrins, which are responsible for the distribution of iron in specific locations; however, serum transferrin (sTf) is the only form that has been proved to have a transport function in the human body. A transferrin consists of one polypeptide chain, which has an N lobe and a C lobe. These lobes are joined to create a central binding site for the iron. The two-lobed structure of transferrin can be seen in Fig. 12.11.

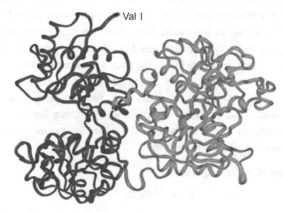

Val 1

■ **FIGURE 12.11** Structure of transferrin.

In the iron-binding site, there are four protein ligands, which include three oxygen anions, one neutral nitrogen atom, as well as a bidentate carbonate ion, resulting in a 6-coordinate complex. The oxygen anions present in this complex are considered to be hard bases due to their small atomic radii and highly negative charges. The three negative charges from these oxygen anions are able to bind to the three positive charges on the Fe^{3+} atom; therefore, sTf is able to bind to two iron atoms, usually in conjunction with two CO_3^{2-} anions. To release the iron from the transferrin protein to be stored or used in the body, one of the lobes of the protein must rotate approximately 55 relative to the other lobe, allowing the iron atoms to exit the binding sites of the protein.

As discussed previously, transferrin only accounts for approximately 0.2% of iron within the body. This is because transferrin is normally only about 30% saturated with iron. This can be explained with an analogy to daily life. The movement of iron in the body is quite similar to the movement of people in a city. At any given time, there is a limited number of people being transported throughout a city on buses or in taxis or cars. This transportation of people is analogous of the transportation of iron; once the people (or iron) reach their destination, they exit the vehicle (or the transferrin).

3.2 **Iron storage: ferritin**

Although transferrin is used to transport iron safely throughout the body, there is a molecule that is responsible for storing excess iron atoms for future use. Ferritin is a protein in the body that is responsible for the storage of iron within cells. Free iron can be toxic, which is why it is necessary to have a location for its safe storage within the body while it is not being utilized or transported by transferrin molecules. Non-protein-bound iron species have the ability to form radicals as shown in Eq. (12.1), which cause cell damage as well as various diseases. Iron-chelating agents found in drugs act to keep the amount of free iron in the body to a minimum, thus having the potential to reduce the incidence of cancer as well as cardiovascular and neurodegenerative diseases. In addition to forming dangerous radicals, iron (III) species are not readily soluble in the human body at a neutral pH, which necessitates that it be bound to a protein to ensure that it does not have adverse effects on the body.

Generation of radical species:

$$Fe(II) + H_2O_2 \rightarrow Fe(III) + {}^{\cdot}OH + OH^- \quad (12.1)$$

Ferritin is shaped like a hollow sphere, with an inner surface lined with carboxylate groups that are able to coordinate up to 4500 Fe^{3+} ions. As can be seen in Fig. 12.12, ferritin exhibits high levels of symmetry, which accounts for its low reactivity and low overall polarity. This low reactivity allows ferritin to be a safe storage place for large quantities of iron atoms.

When an iron atom first enters a ferritin protein through one of the small channels, it is necessary for it to be in a +2 oxidation state. As it enters, it is oxidized to a +3 oxidation state, and then proceeds through reactions as illustrated in Eq. (12.2) to finally be stored as a ferrous-hydric-oxide mineral particle that is then associated with phosphate. The iron in this final storage state is coordinated by oxygen to form a high-spin octahedral structure. These particles have a very unique property. They are considered to be superparamagnetic, which means that they form a large common magnetic moment μ when each particle's magnetic moment is combined.

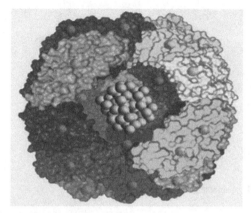

■ **FIGURE 12.12** Structure of ferritin.

Forms of iron as it progresses through the transfer channels to be stored:

$$2Fe^{2+} + O_2 \longrightarrow Fe^{3+}(\mu - O_2)Fe^{3+}$$
$$Fe^{3+}(\mu - O_2)Fe^{3+} + 2H_2O + 2[H] \longrightarrow 2FeO(OH) + 4H^+ \qquad (12.2)$$

3.3 **Siderophores**

Metal ions are important not only in humans, but in bacteria as well. Over 100 enzymes that are active in metabolic processes contain cofactors that utilize iron. Although iron is abundant in the Earth's crust and is heavily relied on for many processes, it is not always readily available to pathogenic bacteria. To regulate the concentration of metal ions and combat this issue, bacteria have developed mechanisms such as efflux systems to reduce metal ion counts, and uptake systems to increase metal ion counts. For example, in many bacteria, one of the most successful and useful uptake systems employs the use of siderophore molecules. These molecules are high-affinity iron chelators, and are used primarily to increase the free iron concentration when necessary (Fig. 12.13).

Siderophores are synthesized in the cytoplasm of the bacteria, and exit the cell to return with bound iron ions. When iron is bound to a siderophore, it is in the ferric form (Fig. 12.14). However, when the siderophore with bound iron reenters the bacteria, the ferric iron is reduced to ferrous iron (Fig. 12.15). These molecules have a higher affinity for iron than does transferrin or lactoferrin, and therefore enable the bacteria to regain needed iron molecules.

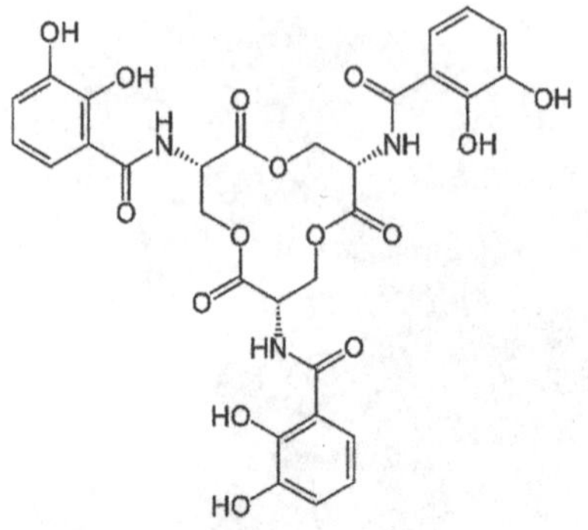

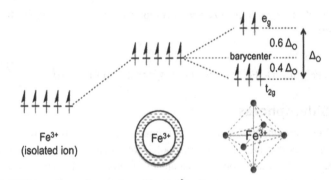

FIGURE 12.13 Enterobactin, a common siderophore.

FIGURE 12.14 Electron diagram of iron (III) with d^5 configuration.

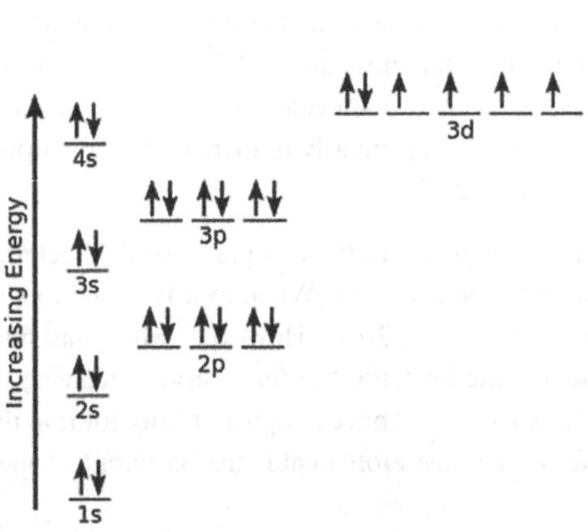

FIGURE 12.15 Electron diagram of iron (II) with d^6 configuration.

As siderophore molecules are hexadentate, when a siderophore molecule binds to an Fe (III) atom, its six donor oxygen atoms coordinate to produce an octahedral structure. Therefore, one Fe (III) atom can bind to one siderophore molecule. When the Fe (III) is bound, it has a d5 configuration and is a high-spin complex, but is typically stable because Fe (III) is a hard acid, while the coordinated oxygen atoms in the siderophore act as hard bases. This is due to the ionic interactions between the hard Lewis acid and hard Lewis base. In contrast, the mild Lewis acid Fe (II) does not readily bind to siderophore molecules due to its preference for atoms such as nitrogen or sulfur, which have weaker hard Lewis base properties.

As stated previously, the balance of transition metals within organisms is extremely important. This applies to humans as well as bacteria. When humans have transition metal concentration abnormalities such as thalassemia or Menkes disease, they become much more susceptible to infection by pathogens. In thalassemia, less hemoglobin is produced due to a low concentration of iron in the body, causing weakness, fatigue, and slow growth. In contrast, Menkes disease is characterized by lower than average amounts of copper in the body, which results in weak muscles, seizures, and low body temperature. In response to infections due to low transition metal concentrations, the immune system begins to distribute transition metals throughout the body. Transferrin and ferritin molecules are active in this process, transporting iron throughout the body to combat the invasion of the pathogen.

3.4 Sodium—potassium pump

Within the body, the transport of metal ions is used to create gradients within cells. One of the most prominent examples of this is the sodium—potassium pump, which is regulated by an enzyme located in the plasma membrane of cells called Na/K-ATPase. This enzyme creates Na^+ and K^+ gradients within cells, which in turn creates a voltage across the cell membrane and a polarity difference between the inside and outside of the cell (Fig. 12.16). This important process uses a large portion of the ATP produced in the body, and creates energy.

The Na^+/K^+-ATPase pumps K^+ out of the cell, which creates a negative polarity inside the cell, and a positive polarity outside the cell. This polarity results in a voltage, which increases or decreases depending on the amount of K^+ being pumped to the outside of the cell. The cell is always trying to reach equilibrium, where the diffusion of K^+ ions out of the cell is equal to the pressure from the external voltage. Each cell has both Na^+ and K^+ channels, which allows ions to enter and exit the cell when the channels are open. At

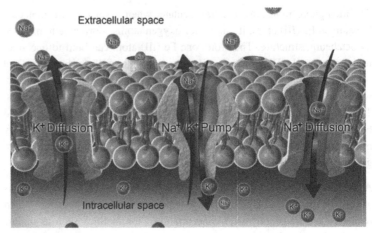

■ FIGURE 12.16 Illustration of the sodium—potassium pump.

rest, K^+ channels are open, resulting in a voltage of 70 mV, with a negative polarity inside the cell and a positive polarity outside the cell. When the Na^+ channels are opened, Na^+ enters the cell, reversing the polarity and changing the voltage. This indicates a signal to the cell.

In addition to maintaining the electrochemical gradient within cells, sodium and potassium ions are used in the nervous system, specifically, within the axons of neurons. When an action potential is sent down the axon, voltage-gated Na^+ channels open, allowing an influx of positive charge, which reduces voltage decay as the signal moves down the axon. After the signal passes one Na^+ channel, a subsequent Na^+ channel opens. Soon after the Na^+ channels begin to open, voltage-gated K^+ channels open as well. This allows an efflux of positive charge, helping the cell potential to return to its standard value of -70 mV. This process plays a very important role in the functioning of humans. Without these channels regulating the charge balances within neutrons, signals would be unable to reach the synapse of the neuron, and neurotransmitters would not be released to the target tissue. The structure of a neuron is shown in Fig. 12.17.

4. OXYGEN TRANSPORT AND ACTIVATION PROTEINS

When they are not being transported or stored themselves, metal ions are aiding with the transportation and storage of other molecules that are imperative to the health of the human body. One of the most prominent examples of this is the function of iron in hemoglobin and myoglobin, which are

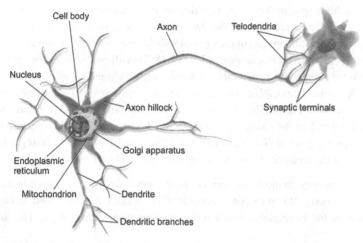

■ **FIGURE 12.17** Structure of a neuron.

biological molecules that are responsible for the transport of oxygen throughout the body, allowing the muscles and tissues to function properly. In other organisms, such as arthropods and mollusks, the hemocyanin molecule is primarily responsible for oxygen transport throughout the organism.

4.1 **Hemoglobin**

One of the most important and prominent uses of transition metals in the body is in the transport of oxygen. Iron, the most abundant transition metal in the body, is a major component of hemoglobin, which is the molecule responsible for oxygen transport through the bloodstream. Hemoglobin is a protein composed of hemes and an active site called an iron porphyrin. Hemoglobin is considered to be a tetramer, as there are four heme molecules in each hemoglobin molecule. Each of the four heme molecules is able to bind to one oxygen molecule, allowing one molecule of hemoglobin to bind to four oxygen molecules. In a heme molecule, porphyrin is a tetradentate ligand found in the center of the heme. This is the location to which the oxygen binds in the hemoglobin.

When hemoglobin is not bound to oxygen, the iron in the heme is in its $+2$ oxidation state resulting in a high-spin complex containing four unpaired electrons. The atom is paramagnetic due to its unpaired electrons, and the diameter of the iron is too large to fit into the plane of the hemoglobin molecule, which therefore is slightly displaced. When hemoglobin has these characteristics, it is considered to be in its "tensed" state, or T form. When oxygen is taken up from the environment, the iron atom is converted from high spin to low spin, resulting in the formation of a diamagnetic molecule and a reduction in the size of the diameter of the iron atom. Specifically, the pi* orbitals of the O_2 molecule interact with the xy and z^2 d orbitals of the iron atom, as seen in Fig. 12.18. This allows the iron atom to fit into the plane of the hemoglobin molecule, and results in the "relaxed" form, or R form of hemoglobin. This change in the conformation of hemoglobin is illustrated in Fig. 12.19. The change in spin state when oxygen is bound is also related to the change in color observed when blood is deoxygenated versus oxygenated. This transition from blue when the blood is deoxygenated to red when the blood is oxygenated can be clearly seen.

Unfortunately, hemoglobin does not bind specifically only to oxygen in the human body. When carbon monoxide (CO) is present, it will bind to the iron in the hemoglobin much more strongly than oxygen does. This is

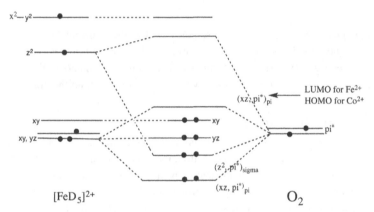

■ **FIGURE 12.18** d-Orbital splitting of heme.

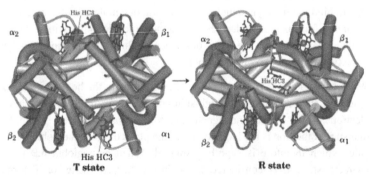

■ **FIGURE 12.19** T state (*left*) and R state (*right*) of hemoglobin.

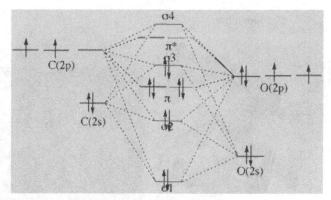

■ **FIGURE 12.20** Molecular orbitals of carbon monoxide.

because the pi* orbitals of CO are empty as seen in Fig. 12.20, and are much more willing to accept electrons from the xy and z^2 orbitals of the iron atom. Because of this increased affinity and bond strength, these molecules remain bound to the hemoglobin, taking up oxygen-binding sites. As an increasing amount of carbon monoxide becomes bound to the hemoglobin, the ability of oxygen to circulate throughout the body is reduced, ultimately causing poisoning and, eventually, death of the individual.

4.2 Myoglobin

After oxygen is transported through the bloodstream by hemoglobin, it is transferred to muscle tissue and stored by myoglobin, which is found in the muscle tissue of vertebrates. Higher concentrations of myoglobin in the muscle tissue allow the muscles to continue to function when oxygen is not actively being inhaled and absorbed into the bloodstream.

This is especially true in mammals such as whales, which spend longer periods of time submerged in water. Myoglobin has a monomeric structure, as seen in Fig. 12.21. It consists of one protein chain and one heme group, and consequently has one binding site for oxygen within the heme group.

The ease with which oxygen is able to be transferred from hemoglobin to myoglobin is primarily due to the differing affinities of these molecules for oxygen. As seen in Fig. 12.22, it is known that myoglobin has a much higher affinity for oxygen until a certain pressure and percent saturation is reached. This is due to the presence of a molecule called 2,3-biphosphoglycerate (BPG) in blood. BPG is highly negatively charged, and is present in deoxygenated hemoglobin. It reduces the ability of the iron atom to move into the plane of the hemoglobin because of steric effects,

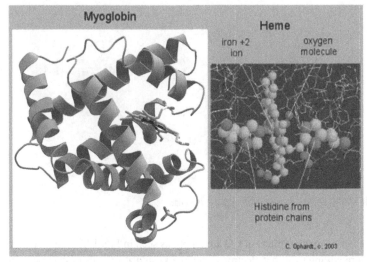

■ **FIGURE 12.21** Myoglobin structure.

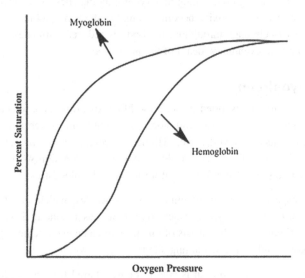

■ **FIGURE 12.22** Affinity of hemoglobin and myoglobin for oxygen.

consequently lowering the oxygen affinity. The structure of myoglobin is more open and allows the iron atom to move more freely, which enables the iron atom to bind more easily with incoming oxygen. This allows the oxygen to be easily transferred from the bloodstream to the muscle tissue throughout the body.

4.3 **Hemocyanin**

In comparison with hemoglobin and myoglobin, hemocyanin has a similar function, but exhibits a significantly different structure and mechanism of oxygen transport. Hemocyanin molecules are found free in the blood of arthropods and mollusks, and are the main method of oxygen transport within these organisms. In contrast to hemoglobin, hemocyanin utilizes two copper atoms at the binding site to bind to a total of up to 160 O_2 molecules, with one dioxygen atom per two copper atoms. Although there are structural differences in the hemocyanin molecules found in arthropods in comparison with the hemocyanin molecules found in mollusks, they all have a very similar mechanism at the active site. When no oxygen is bound to the hemocyanin, it is termed deoxyhemocyanin. When oxygen becomes associated with the active site, it is termed oxyhemocyanin. These forms can be seen in Fig. 12.23.

This molecule, in the deoxygenated state, consists of two Cu(I) cations, which coordinate to one dioxygen molecule. In examining the mechanism of oxygen binding to hemocyanin, it has been discovered that the valence

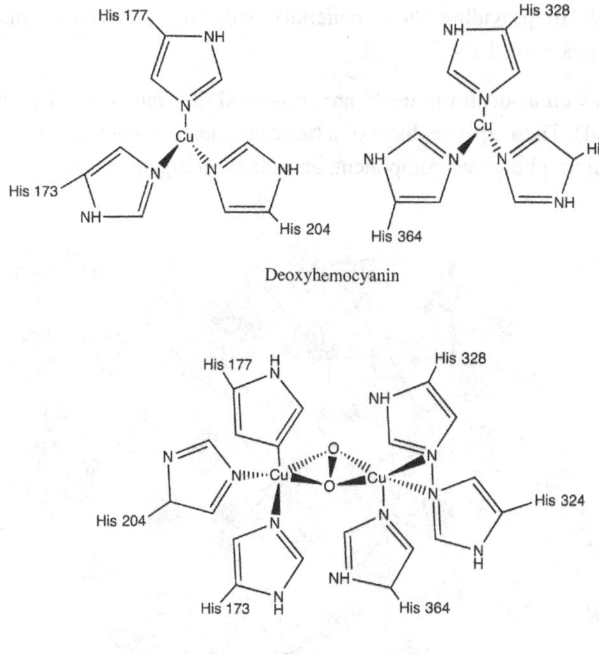

Deoxyhemocyanin

Oxyhemocyanin

■ **FIGURE 12.23** Deoxyhemocyanin and oxyhemocyanin.

state of the copper atoms does not simply switch between Cu(I) and Cu(II), as was thought previously. It has been suggested that half of the metal is in the +1 oxidation state, and the other half is in the +2 oxidation state. This is due to the participation of water molecules in the process. When hemocyanin is in its deoxygenated state, water can partially replace the oxygen molecules, which withdraws d-orbital electrons from the copper atoms.

5.   **BIOMINERALIZATION**

In addition to their importance within the body, inorganic molecules also function to provide organisms with mechanical strength. Biomineralization, the controlled formation of inorganic materials containing calcium or silicon in a living body, is the basis for the formation of bones, teeth, and outer shells, which provide structure and strength. Some of the most common components of these structures include calcium phosphate in bones and dentin, and calcium carbonate in shells. In addition, silica is utilized in the formation of a protective structure found surrounding diatoms called the frustule.

The unique properties of calcium make it extremely useful in structural materials. The ability of the Ca(II) cation to form bridges between two negatively charged molecules enables the formation of very strong bonds, which aids in providing these materials with strength without making them too heavy or dense.

Bones, as well as dentin in teeth, are composed of a mineral called apatite (Fig. 12.24). These apatites have of a basic composition containing an inorganic calcium phosphate component, as well as an organic component. The

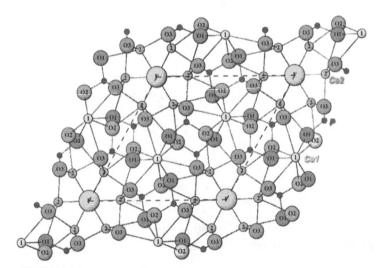

■ **FIGURE 12.24** Apatite structure of bone.

inorganic component is responsible for structural integrity and the material's ability to withstand pressure, while the organic component gives it flexibility and increases its resistance to fracture. The calcium phosphate component found in bone accounts for approximately 66% of its total weight, which makes it extremely strong and allows bone to bear weight. Although the general structure of apatite is similar for all structural materials, different ions may appear in different materials based on what they are used for. For example, in bone formation, osteoblasts are responsible for forming a matrix with phosphoproteins, which allows Ca(II) cations to form strong bonds at regular intervals. Small gaps within the nanocrystalline structure of bone contain free Ca^{2+} and PO_4^{3-} ions. These ions are used to replace dissolved ions in the structure, which aid in maintaining the structural soundness of bone. This material is commonly called hydroxylapatite, which has a composition of $Ca_5(PO_4)_3OH$. The biological apatite found in bone and other structural materials also contains a relatively small number of Mg^{2+}, Na^+, K^+, Cl^-, and F^- ions, as well as some CO_3^{2-} ions.

Enamel, which composes the outermost layer of teeth, comprises a material that is approximately 90% inorganic and contains no living cells. This material is extremely strong and hard due to its crystalline structure, but cannot be reproduced or replaced. By using dental products containing fluoride, a material called fluorapatite is created on the surface of the enamel, which helps to increase the longevity of the enamel. Although apatites are generally beneficial, problems can arise when toxic metal ions such as Pb^{2+} and As^{5+} are incorporated into its structure. Unfortunately, this can happen easily due to the nature of apatite to easily incorporate different ions into its structure. For example, when lead pipes are used to carry water, the lead will form lead phosphate, which is insoluble and remains in the water in an apatitic structure, causing increased water Pb levels.

BIBLIOGRAPHY

1. Rehder, D. *Introduction to Bioinorganic Chemistry;* University of Lund, 2008.
2. Bertini, I.; Gray, H.; Lippard, S.; Valentine, J. *Bioinorganic Chemistry;* University Science Books: Mill Valley, CA, 1994.
3. Kaim, W.; Schwederski, B.; Klein, A. *Bioinorganic Chemistry - Inorganic Elements in the Chemistry of Life: An Introduction and Guide,* 2nd ed.; John Wiley & Sons, 2013. s.l., [Online].
4. Mjos, K. D.; Orvig, C. Metallodrugs in Medicinal Inorganic Chemistry. *Chem. Rev.* **2014,** *114,* 4540−4563.
5. Hartinger, C. G.; Metzler-Nolte, N.; Dyson, P. J. Challenges and Opportunities in the Development of Organometallic Anticancer Drugs. *Organometallics* **2012,** *31,* 5677−5685.

6. Lloyd, N. C.; Morgan, H. W.; Nicholson, B. K.; Ronimus, R. S. The Composition of Ehrlich's Salvarsan: Resolution of a Century-old Debate. *Angew. Chem. Int. Ed. Engl.* **2005,** *44,* 941−944.

7. Abeles, R. H.; Dolphin, D. The Vitamin B12 Coenzyme. *Acc. Chem. Res. [Online]* **1976,** 114−120.

8. Medek, A.; Frydman, L. *J. Am. Chem. Soc. [Online]* **2000,** *122* (4).

9. Matthews, R. G. Cobalamin-dependent Methyltransferases. *Acc. Chem. Res. [Online]* **2001,** *34* (8).

10. Use of Cisplatin for Cancer Treatment. http://www.cancer.gov/research/progress/discovery/cisplatin.

11. Wang, D.; Lippard, S. J. *Nat. Rev. Drug Discov.* **2005,** *4,* 307.

12. Sresht, V.; Bellare, J. R.; Gupta, S. K. Modeling the Cytotoxicity of Cisplatin. *Ind. Eng. Chem. Res. [Online]* **2011,** *50* (23), 12872−12880.

13. Wheate, N. J.; Walker, S.; Craig, G. E.; Oun, R. The Status of Platinum Anticancer Drugs in the Clinic and in Clinical Trials. *Dalton Trans.* **2012,** *39,* 8113−8127.

14. Chellan, P.; Sadler, P. J. The Elements of Life and Medicines. *Phil. Trans. R. Soc. A* **2015,** *373,* 20140182.

15. Köpf, H.; Köpf-Maier, P. *Angew. Chem.* **1979,** *91,* 509.

16. Kröger, N.; Kleeberg, U. R.; Mross, K.; Edler, L.; Hossfeld, D. K. *Onkologie* **2000,** *23,* 60−62.

17. Köpf-Maier, P.; Köpf, H.; Neuse, E. W. *Angew. Chem.* **1984,** *96,* 446−447.

18. Dubar, F.; Khalife, J.; Brocard, J.; Dive, D.; Biot, C. Ferroquine, an Ingenious Antimalarial Drug: Thoughts on the Mechanism of Action. *Molecules* **2008,** *13,* 2900−2907.

19. Leaper, D. J. Silver Dressings: Their Role in Wound Management. *Int. Wound J.* **2006,** *3,* 282−294.

20. Klasen, H. J. Historical Review of the Use of Silver in the Treatment of Burns. I. Early Uses. *Burns* **2000,** *26,* 117−130.

21. Kean, W. F.; Kean, I. R. L. Clinical Pharmacology of Gold. *Inflammopharmacology* **2008,** *16* (3), 112−125.

22. Raubenheimer, H. G.; Schmidbaur, H. The Late Start and Amazing Upswing in Gold Chemistry. *J. Chem. Educ.* **2014,** *91,* 2024−2036.

23. Mayo Clinic. http://www.mayoclinic.org/tests-procedures/mri/.

24. Callaghan, P. *Principles of Nuclear Magnetic Resonance Microscopy,* Revised ed.; Oxford Press: Oxford, 1994.

25. Raymond, K. N.; Pierre, V. C. Next Generation, High Relaxivity Gadolinium MRI Agents. *Bioconjugate Chem.* **2005,** *16,* 3−8.

26. Li, Y.; Beija, M.; Laurent, S.; vander Elst, L.; Muller, R.; Duong, T. T.; Lowe, A. B.; Davis, T. P.; Boyer, C. Macromolecular Ligands for Gadolinium MRI Contrast Agents. *Macromolecules* **2012,** *45,* 4196−4204.

27. Caravan, P.; Ellison, J. J.; McMurry, T. J.; Lauffer, R. B. Gadolinium(III) Chelates as MRI Contrast Agents: Structure, Dynamics, and Applications. *Chem. Rev.* **1999,** *99,* 2293−2352.

28. Wang, Y. J. Superparamagnetic Iron Oxide Based MRI Contrast Agents: Current Status of Clinical Application. *Quant. Imaging Med. Surg.* **2011,** *1* (1), 35−40.

29. Bourassa, M. W.; Miller, L. M. *Metallomics* **2012,** *4,* 721.

30. Banerjee, S.; Raghavan, M.; Pillai, M. R. A.; Ramamoorthy, N. *Semin. Nucl. Med.* **2001,** *31,* 260.

31. World Nuclear Association. Radioisotopes in Medicine. http://www.world-nuclear. org.

32. Liu, S.; Edwards, D. S. *Chem. Rev.* **1999,** *99,* 2235.

33. Baker, H. M.; Anderson, B. F.; Baker, E. N. Dealing with Iron: Common Structural Principles in Proteins that Transport Iron and Heme. *Proc. Natl. Acad. Sci. [Online]* **2003,** *100,* 3579–3583.

34. Hamilton, D. H.; Battin, E. E.; Lawhon, A.; Brumaghim, J. L. Using Proteins in a Bioinorganic Laboratory Experiment: Iron Loading and Removal from Transferrin. *J. Chem. Educ. [Online]* **2009,** *86,* 969.

35. Frankel, R. B.; Papaefthymiou, G. C.; Watt, G. D. Variation of Superparamagnetic Properties with Iron Loading in Mammalian Ferritin. *Hyperfine Interact. [Online]* **1991,** *66,* 71–82.

36. Pankhurst, Q. A.; Connolly, J.; Jones, S. K.; Dobson, J. Applications of Magnetic Nanoparticles in Biomedicine. *J. Phys. D. Appl. Phys. [Online]* **2003,** 36.

37. Miethke, M.; Marahiel, M. A. Siderophore-based Iron Acquisition and Pathogen Control. *Microbiol. Mol. Biol. Rev. [Online]* **2007,** 413–451.

38. Porcheron, G.; Garénaux, A.; Proulx, J.; Sabri, M.; Dozois, C. M. Iron, Copper, Zinc, and Manganese Transport and Regulation in Pathogenic Enterobacteria: Correlations between Strains, Site of Infection and the Relative Importance of the Different Metal Transport Systems for Virulence. *Front. Cell. Infect. Microbiol. [Online]* **2013,** *3* (1).

39. Apell, H.-J.; Benz, G.; Sauerbrunn, D. *Biochemistry. [Online]* **2011,** *50* (3), 409–418.

40. Silverthorn, D. U.; Johnson, B. R. *Human Physiology,* 7th ed.; Pearson, 2016.

41. Ophardt, C. Globular Proteins. Elmhurst College Chemistry. http://chemistry. elmhurst.edu/vchembook/568globularprotein.html.

42. Hemoglobin. Stanford University. http://web.stanford.edu/~kaleeg/chem32/hemo/.

43. Boyer, R. Interactive Concepts in Biochemistry. Interactive Concepts in Biochemistry - Structure Tutorials. http://www.wiley.com/college/boyer/0470003790/structure/hbmb/ eval.html.

44. Panzer, D.; Beck, C.; Hahn, M.; Maul, J.; Schonhense, G.; Decker, H.; Aziz, E. F. Water Influences on the Copper Active Site in Hemocyanin. *J. Phys. Chem. Lett. [Online]* **2010,** *1,* 1642–1647. pubs.acs.org/JPCL.

45. Senozan, N. M. Hemocyanin: The Copper Blood. *J. Chem. Educ. [Online]* **1976,** *53* (11).

46. Ochial, E.-I. Biomineralization. *J. Chem. Educ. [Online]* **1991,** *68* (8), 627–630.

47. Vallet-Regi, M.; Navarrete, D. A. *Nanoceramics in Clinical Use: From Materials to Applications,* 2nd ed.;, 2015, 2, 1–29.

48. Wopenka, B.; Pasteris, J. A Mineralogical Perspective on the Apatite in Bone. *Mater. Sci. Eng. [Online]* **2005,** 131–143.

Index

'*Note*: Page numbers followed by "f" indicate figures and "t" indicate tables.'

A

Abzymes, 209
Alkene complexes, 205–207,
    205f–206f
Anionic complexes, 82–83
Arene complexes, 107–111, 108f–111f
Aromatic complexes, 207, 207f
Auranofin, 232, 232f

B

Back π bonding, 78
Base hydrolysis, 140–143, 143t
Bent geometries, 20f
Benzene sandwich complexes,
    112f–114, 113f–114f
BF_3, 61–63, 62f
Bidentate ligands, 79
Bioinorganic chemistry/applications, 225
    biomineralization, 246–247, 246f
    cisplatin and cancer treatments, 228
        anticancer platinum metallodrugs
            carboplatin, 229, 229f
            ferrocifen, 230–231, 231f
            formation, 228–229
            K1019, 230, 231f
            (N)KP1339, 230, 231f
            NAMI-A, 230, 231f
            square planar structure, 228, 228f
            titanocene chloride, 230, 230f
        diagnostic metallodrugs, 233–234,
            233f
    metal ions, transport and storage,
        234–235
        iron storage, 235–237, 235f
        siderophores, 237–239, 238f
        sodium-potassium pump, 239–240,
            240f–241f
    oxygen transport and activation
        proteins, 240–241
        hemocyanin, 245–246, 245f
        hemoglobin, 241–243, 242f–243f
        myoglobin, 243–244, 244f
    salvarsan, 226, 226f
    therapeutic applications, of organometallic
        compounds, 231–232, 232f
    vitamin B12, 227, 227f

Black-body radiation, 3
π bonding, 104–107, 105f
σ bonding, 53–55
Bonding theories
    crystal field theory, 92–99
        distortions, 98–99
        experimental evidence stabilization,
            94, 94f
        geometries complexes, 95–96
        octahedral complexes, 92–94, 93f,
            94t, 97–98, 98t
        predictions, 97–98
        spin pairing complexes, 97–98
        square planar complexes, 96, 96f, 98
        stabilization energy, 96–99, 97t
        tetrahedral complexes, 95, 95f, 98
        trigonal bipyramidal field, 96
    metal complexes, 89
    molecular orbital (MO) theory,
        99–114
        arene complexes, 107–111,
            108f–111f
        arene-like ligands, 112, 112f
        benzene sandwich complexes,
            112–114, 113f–114f
        π bonding, 104–107, 105f
        effective atomic number (EAN) rule,
            106–107, 106f–107f
        octahedral complexes, 100–102,
            100f–101f, 101t, 104–107, 105f
        square planar complexes,
            104, 104f
        tetrahedral complexes, 102–103,
            102f, 102t, 103f
    valence bond theory, 90–92
        coordination compounds, 90
        coordination number four, 91–92
        coordination number six, 90–91
Bridging carbonyl, 203–205
Bridging complexes, 83

C

Catalytic olefin metathesis, 220, 221f
Catalytic processes
    acetic acid from CH_3OH, 216–217,
        216f

alkene polymerization with Ziegler-
    Natta catalyst, 217–218
defined, 209
heterogeneous catalysis, 217–218
homogeneous catalysts, 209–211
hydrogenation catalysts, 212–215
olefin metathesis with Grubbs/Schrock
    catalysts, 220
Sonogashira cross-coupling reaction, 218
Suzuki–Miyaura cross-coupling
    reaction, 218
Center of inversion, 36
CFSE. *See* Crystal field stabilization
    energy (CFSE)
Characters of matrices, 49
Character tables, 50–51, 50t
Charge transfer spectra, 168–170, 169f
Chlorophyll, 76
Cisplatin/cancer treatments, 228
    anticancer platinum metallodrugs
        carboplatin, 229, 229f
        ferrocifen, 230–231, 231f
        formation, 228–229
        K1019, 230, 231f
        (N)KP1339, 230, 231f
        NAMI-A, 230, 231f
        square planar structure, 228, 228f
        titanocene chloride, 230, 230f
Cluster chemistry
    boranes, 172–179
    boron clusters bonding, 174–176,
        174f–176f
    carboranes, 172–179
    overview, 171
    polyhedral boron clusters, 172, 172f–173f
    Wade's rules, 176–179, 177f,
        177t–178t, 179f, 179t, 180f,
        182–194, 183f, 183t
        capping groups, 185–186, 186f
        condensed clusters, 186–187,
            186f–187f
        elements and transition metals,
            180–182
        interstitial atoms, 188
        isolobal relationships, 188–193,
            188f–189f, 189t, 190f–193f

Cluster chemistry (*Continued*)
  mixed main group/transition metal
    clusters, 184e–185, 184f, 185t
  noble gas structures, 193–194, 193f
  Zintl anions, 180–182, 181f
Cobalt carbonyl, 215
Complementary color, 89
Complex ions, 75–76
Coordination chemistry, 199
  activation parameters
    activation volume, 121
    enthalpy and entropy of activation,
      119–121
    mechanistic studies, 121–123, 122t
  coordination numbers reactions,
      123–149
    base hydrolysis, 140–143, 143t
    electron transfer reactions, 145–149
    five-coordinate complexes, 127–129,
      129f
    four-coordinate complexes,
      126–127
    group 13 tetrahedral complexes, 126,
      126t
    group 14 tetrahedral complexes,
      126–127
    heavier group 9 metals, 144,
      144t–145t
    inner sphere mechanism, 145–147,
      146f
    isomerization during substitution,
      139–140
    kinetic effect, 133, 133f
    outer sphere mechanisms,
      147–149
    phosphorus, 127–129, 127f
    six-coordinate octahedral complexes,
      134–139, 135t, 137t, 138f, 138t,
      139f, 139t
    square planar complexes, 129–132,
      130t–131t
    sulfur, 127–129
    three-coordinate complexes,
      123–125, 124t–125t
    trans effect, 132–133, 132f
    two- to six-coordinate complexes,
      123, 123t
  metal complexes complications
    complex ligands, 119, 120t
    solvent competition, 119
  overview, 115

substitution reaction mechanisms,
    116–118
  associative mode, 116–117, 116f
  dissociative mode, 117–118, 118f
Coordination compounds
  International Union of Pure and
    Applied Chemists (IUPAC) rules,
    80–81
  nomenclature, 80–83
Coordination number = 4, 84–87,
    84f–86f, 87f
Coordination sphere, 115
Core electrons, 7
$Cr(CO)_6$, 107
Crystal field stabilization energy
    (CFSE), 96–99, 97t
Crystal field theory, 92–99
  crystal field stabilization energy
    (CFSE), 96–99, 97t
  distortions, 98–99
  experimental evidence stabilization, 94,
    94f
  geometries complexes, 95–96
  octahedral complexes, 92–94, 93f, 94t,
    97–98, 98t
  predictions, 97–98
  spin pairing complexes, 97–98
  square planar complexes, 96, 96f, 98
  tetrahedral complexes, 95, 95f, 98
  trigonal bipyramidal field, 96

D

Direct product, 52

E

Effective atomic number (EAN) rule,
    106–107, 106f–107f, 201
Electron–electron repulsion, 7–8
Electronic spectra
  charge transfer spectra, 168–170, 169f
  La Porte selection rule, 156
  molar absorptivity, 156–157
  octahedral (O_h) complexes, 157–158,
    157t
  Orgel diagrams, 158–167, 158f
    distortion effect, 160, 160f
    F states systems, 160–167,
      161f–162f, 165f–166f
    simple one-electron approach, 159
  overview, 155
  selection rules, 155–157

spin selection rule, 156
Tanabe–Sugano diagrams, 167, 168f
tetrahedral (T_d) complexes, 157–158,
    157t
vibronic coupling, 156
Electron spin–intrinsic properties, 5–6
Electron transfer reactions, 145–149

F

$Fe(CO)_5$, 107
Five-coordinate complexes, 127–129,
    129f
Four-coordinate complexes, 126–127

G

Gouy balance, 89
Group symmetry orbitals, 52
Group theory
  cyclopentadienide ion, 64, 64f
  molecular spectroscopy
    symmetry normal modes, 65–69,
      66f, 67t, 68f
    types, 65
    vibration normal modes, 65–69, 66f,
      67t, 68f
  molecular structure
    BF_3, 61–63, 62f
    molecular orbital correlation dia-
      gram, 61–63
    quantum mechanics, application to,
      51–53
    triangular planar structure, 53–61,
      54f–55f, 56t, 57f–61f
  π orbitals, 64
  overview, 43
  square planar ML_4, 63–64
  symmetry operations properties,
      43–51
    character tables, 50–51, 50t
    irreducible representations, 48–50,
      50t
    reducible representations, 48–50, 50t
    representations, 46–48, 46t, 47f
    sequential operations, 43–46,
      44f–45f, 45t

H

Heavier group 9 metals, 144, 144t–145t
Heme group, 75
Heterogeneous catalysis, 217–218
Homogeneous catalysis, 76

insertion reaction, 211
oxidative addition and reductive
    elimination, 209—211, 210f
Hund's rules, 11
Hydroformylation, 214—215, 215f
Hydrogenation catalysts
    hydroformylation, 214—215, 215f
    monohydride complexes, 213, 214f
    types, 212
    Wilkinson's catalyst
        characteristics, 213
        higher coordinated solvo-complex
            formation, 213, 213f
        mechanism, 212, 212f

I

Improper axes of rotation, 36
Infrared (IR) active, 68
Inner sphere mechanism, 145—147,
    146f
Insertion reaction, 211
International Union of Pure and Applied
    Chemists (IUPAC) rule, 80—81
Iron storage
    ferritin, 236—237, 237f
    transferrin, 235—236, 235f
Irreducible representations, 48—50, 50t
Isomer, 37
Isomerism, 84—87
Isomerization during substitution,
    139—140

J

Jahn—Teller distortion, 99

K

Kinetic effect, 133, 133f

L

Lewis acid, 77
Ligands and metal complexes
    coordination compounds
        International Union of Pure and
            Applied Chemists (IUPAC) rules,
            80—81
        nomenclature, 80—83
        isomerism, 84—87
        coordination number = 4, 84—87,
            84f—86f, 87f
    overview, 75—76
    transition metals, 76—80

coordination compounds, 77—78
electronic structures, 76—77, 76t
Lewis bases, 78—80, 80f
ligands, 78—80, 79f
molecular structure, 78f
oxidation states, 76—77, 77t
Linear geometries, 21f

M

Magnetic moment, 89
Magnetic properties, 5
Magnetic susceptibility, 89
Many electron atom, 6—8
Matrix representations, 46—48
Metal carbonyls, 205t
    $Co_2(CO)_8$, electronic arrangement/
        possible two structures, 202—205,
        203f—204f
    $Fe(CO)_5$, electronic arrangement of,
        202, 202f
    $Mn_2(CO)_{10}$, electronic arrangement/
        structure of, 203—205, 203f
    multimetal carbonyl complexes,
        possible isomeric structures,
        203—205, 204f
    $Ni(CO)_4$, electronic arrangement of,
        202, 202f
Mirror planes, 36
Molar absorptivity, 156—157
Molecular geometries
    nonrigid shapes of molecules
        general concept, 26
        specific examples, 27—30
    overview, 15
    shapes of molecules
        other considerations, 25—26
        specific examples, 19—25, 19f
        valence shell electron pair repulsion
            (VSEPR) model, 16—18
Molecular motion, 65
Molecular orbital correlation diagram,
    61—63
Molecular orbital (MO) theory, 99—114
    arene complexes, 107—111, 108f—111f
    arene-like ligands, 112, 112f
    benzene sandwich complexes,
        112—114, 113f—114f
    π bonding, 104—107, 105f
    effective atomic number (EAN) rule,
        106—107, 106f—107f
    octahedral complexes, 100—102,
        100f—101f, 101t, 104—107, 105f

square planar complexes, 104, 104f
tetrahedral complexes, 102—103, 102f,
    102t, 103f
Molecular spectroscopy
    symmetry normal modes, 65—69, 66f,
        67f, 68f
    types, 65
    vibration normal modes, 65—69, 66f,
        67t, 68f
Molecular structure
    BF_3, 61—63, 62f
    molecular orbital correlation diagram,
        61—63
    quantum mechanics, application to,
        51—53
    triangular planar structure, 53—61,
        54f—55f, 56t, 57f—61f
Molecular symmetry
    elements, 32—34
        operations, 31—32, 32f
        symmetry operations, 31—32
    examples, 35—38
    operations, 32—34
    overview, 31
    PBr_5, 35, 35f
    PF_2Cl_3, 37, 37f
    point groups, 38—42
        assigning rules, 39
        examples, 39—42, 40t
        symmetry, 41t
Monoclonal antibodies, 209
Monodentate ligands, 78
Monohydride complexes, 213, 214f

N

Naming complex cations,
    81—83
Neutral complexes, 81—83
n-fold improper axis of rotation, 34
Niels Bohr's model, 3
Normal modes, 65—69

O

Octahedral complexes, 18, 18f, 22f,
    85—87, 100—102, 100f—101f,
    101t, 104—107, 105f, 157—158,
    157t
Olefin metathesis, 220, 221f
Order of the group, 46
Organometallic chemistry, 199
Organometallic compounds, 199

Organometallic compounds (*Continued*)
  alkene complexes, 205–207,
    205f–206f
  aromatic complexes, 207, 207f
  effective atomic number rule, 201
  IUPAC nomenclature, 201, 201f
  metal carbonyls, 205t
    $Co_2(CO)_8$, electronic arrangement/
      possible two structures, 202–205,
      203f–204f
    $Fe(CO)_5$, electronic arrangement of,
      202, 202f
    $Mn_2(CO)_{10}$, electronic arrangement/
      structure of, 203–205, 203f
    multimetal carbonyl complexes,
      possible isomeric structures,
      203–205, 204f
    $Ni(CO)_4$, electronic arrangement of,
      202, 202f
  therapeutic applications, 231–232,
    232f
  types of, 200, 200f
Orgel diagrams, 158–167, 158f
  distortion effect, 160, 160f
  F states systems, 160–167, 161f–162f,
    165f–166f
  simple one-electron approach, 159
Outer sphere mechanisms,
    147–149
Oxidative addition/reductive
    elimination, 209–211, 210f

P

Pauli's exclusion principle, 9
Penta- and hexadentate ligands, 80
Phosphorus, 127–129, 127f
Polydentate ligands, 78–79
Polytopal isomerization, 28
Projection formula, 53
Pseudorotation mechanism, 28f
Pyrrole groups, 75

Q

Quantum mechanics, 51–53
Quantum theory
  hydrogen atom, 3–13
    electron–electron repulsion, 7–8
    many electron atom, 6–8
    quantum numbers and significance,
      4–6, 5f, 6t

spin–orbit coupling, 11–13, 12f
valence–valence repulsion, 8–11
overview, 3

R

Reducible representations, 48–50, 50t
Reduction equation, 56
Reduction formula, 49–50
Representations, 46–48, 46t, 47f
Rotational motion, 65

S

SALC. *See* Symmetry Adapted Linear
    Combinations (SALC)
Salvarsan, 226, 226f
Sandwich compounds, 207
Schrock catalysts, 220, 221f
Seesaw, 21f
Sequential operations, 43–46, 44f–45f,
    45t
Siderophores, 237–239, 238f
Silver sulfadiazine, 232, 232f
Six-coordinate octahedral complexes,
    134–139, 135t, 137t, 138f,
    138t, 139f, 139t
Sodium-potassium pump, 239–240,
    240f–241f
Sonogashira cross-coupling reaction,
    218, 219f
Spin–orbit coupling, 11–13, 12f
Square planar complexes, 84–85, 104,
    104f, 129–132, 130t–131t
Square planar geometries, 22f
Square planar ML_4, 63–64
Square pyramid, 22f
Sulfur, 127–129
Suzuki-Sonogashira coupling, 199, 218,
    219f
Symmetry Adapted Linear
    Combinations (SALC), 52
Symmetry normal modes, 65–69, 66f,
    67t, 68f
Symmetry operations properties, 43–51
  character tables, 50–51, 50t
  irreducible representations, 48–50, 50t
  reducible representations, 48–50, 50t
  representations, 46–48, 46t, 47f
  sequential operations, 43–46, 44f–45f,
    45t
Syphilis, 226

T

Tanabe–Sugano diagrams, 167, 168f
Term symbols, 10
Tetradentate ligands, 79–80
Tetrahedral complexes, 17, 17f, 20f, 84,
    102–103, 102f, 102t, 103f,
    157–158, 157t
Three-coordinate complexes, 123–125,
    124t–125t
Titanocene chloride, 230, 230f
Total orbital angular momentum, 8–9
Total spin angular momentum,
    9–10
Trans effect, 132–133, 132f
Transferrin, 235–236, 235f
Transition metal complex, 115
Transition metals, 76–80
  coordination compounds,
    77–78
  electronic structures, 76–77, 76t
  Lewis bases, 78–80, 80f
  ligands, 78–80, 79f
  molecular structure, 78f
  oxidation states, 76–77, 77t
Triangular planar structure, 53–61,
    54f–55f, 56t, 57f–61f
Tridentate ligands, 79
Trigonal bipyramidal, 17, 17f, 21f
Trigonal planar, 16, 16f
Trigonal prism, 18, 18f
Trigonal pyramid, 20f
Tripodal ligand, 79–80
T-shaped, 21f
Two- to six-coordinate complexes, 123,
    123t

V

Valence bond theory, 90–92
  coordination compounds, 90
  coordination number four, 91–92
  coordination number six, 90–91
Valence electrons, 7
Valence shell electron pair repulsion
    (VSEPR) model, 16–18
Valence–valence repulsion, 8–11
Vibrational motion, 65
Vibration normal modes, 65–69, 66f,
    67t, 68f
Vibronic coupling, 156
Vitamin B12, 76, 227, 227f

VSEPR model. *See* Valence shell electron pair repulsion (VSEPR) model

W

Wade's rules, 176—179, 177f, 177t—178t, 179f, 179t, 180f, 182—194, 183f, 183t
  capping groups, 185—186, 186f
  condensed clusters, 186—187, 186f—187f
elements and transition metals, 180—182
interstitial atoms, 188
isolobal relationships, 188—193, 188f—189f, 189t, 190f—193f
mixed main group/transition metal clusters, 184—185, 184f, 185t
noble gas structures, 193—194, 193f
Wilkinson's catalyst characteristics, 213
higher coordinated solvo-complex formation, 213, 213f
mechanism, 212, 212f

Z

Zero-order approximation, 6—7
Ziegler—Natta catalyst, 217—218, 217f—218f
Ziegler—Natta polymerization, 199
Zintl anions, 180—182, 181f

Printed in the United States
By Bookmasters